On The Battlefield

A MEMOIR OF AN ETHIOPIAN AIRBORNE SOLDIER

Tesfaye Habtemariam, Brigadier General

Translation by
Mekonnen Haile
and
Sahlesellassie Berhanemariam

ISBN-13: 978-0-692-18126-3
ISBN-10: 0-692-18126-1

contact@battlefieldmemoirs.com

Acknowledgments

I am indebted to the following people for their support and assistance with the translation of my book:

My late wife, Fessessework Tessema, who supported me at every stage when I was writing my book in Amharic.

My daughter, Awetash Tesfaye who typed the English manuscript.

My granddaughter, Mariamawit Amare who drew the maps in the book, did photographic editing and provided valuable assistance in the preparation of the images.

Amare Tesfaye, Nako Amare, Mulugeta Kebede, Maged Sharabi, Dereje Demissie, Darren Gilbert who proofread the translated manuscript and gave valuable comments and edits.

And last but not least, to all my children and grandchildren, family and friends who have been supportive over the course of the years and whose names I have not mentioned.

Contents

Introduction

Over the years, many friends and former brothers-in-arms who knew of my history with the Ethiopian Army, have asked me to share my story of service and sacrifice with the Ethiopian people. Typically I answered, "I need some time to think about it." I did feel that I should share my battlefield experiences in defense of my country with the Ethiopian people. But there were several issues I first had to work through.

The primary challenge was my conviction that when presenting such a story to the public, it needed to be told accurately and understandably. I also wanted it to be exciting and enjoyable to read. However, since I had no experience as a writer, I was not sure how to begin a project such as this book.

Another challenge was that the battles and events I would be describing took place more than twenty years in the past. Many witnesses and former colleagues of mine were dead. Some were unavailable, and some were even in exile.

Most of the research and documentation for a project such as this was no longer easily accessible. Things such as military maps, records, telegrams, and operation and intelligence documents were not easy to get. Many of my private notes were scattered and lost over the years as

I moved from place to place on various assignments and tours of duty. I just did not know where to begin.

In 2002, my son Amare invited me to visit him in Botswana, where he resided at the time. During my stay in Botswana, my son would ask me to tell him stories of my experiences in the Ethiopian Army and on the battlefield. We spent many evenings together as I described for him my time in service to my country.

Later, he suggested I write these stories down. Therefore, I began to write while he and his wife were at work. It quickly became a labor of love, and I made good progress.

The process of thinking back and revisiting old memories of my battle experiences had a positive effect on me. I worked hard to remember the events and facts, placing them in proper chronological order. I enjoyed revisiting the memories of my brothers-in-arms and the men I commanded. Memories of weapons and tactics, the various battlefields came streaming back to me. I also reflected upon the terrible cost in human lives and material. Day by day, my ability to recall these events improved, helping me to put together the past to complete this project. When my visit to Botswana was coming to an end, and I prepared to return home to Addis, my son strongly urged me to continue my writing. I decided I would.

As soon as I returned home, I went to work in earnest. I gathered up my notebooks and began serious research. At the Ministry of Defense, I was able to access some documents predating the Ethiopian People's Revolutionary Democratic Front which was related to some of the military actions in which I was involved. I found old battlefield maps and was able to interview military personnel of all ranks. Their support was very helpful to me in providing a strong foundation for my project.

My original plan was to focus on the defensive battle at Nakfa (a town in Eritrea). As my research progressed, I found that to be too narrow a scope. I decided to widen the extent of my story to include the eastern and northern battlefronts as well.

The eastern and northern battlefronts saw a vast array of modern weapons and the sacrifice of many tens of thousands of men. Millions of ETB (Ethiopian currency) worth of property was destroyed. These battles have contributed valuable lessons in military theory and science, and we should examine them from many different angles to fully understand any lessons to be learned. Military leaders and experts from all three branches of the armed forces (Air Force, Army, and Navy) must tell the story. Only a collective effort such as this can ensure these battles receive the true significance they deserve.

For this book, I have selected battles in which I was physically present and given command. I have made a real effort to portray an accurate and objective account of these events, giving both sides in order to provide the benefit of an intellectually honest story.

CHAPTER 1
Childhood

My name is Tesfaye Habtemariam, and I was born in Sise Village, located in the Shoa province of Ethiopia. My father was called Quotiye Gana (Habtemariam was his Christian name), and my mother was called Fershiwot Gajwe. I am the third born of nine children. Since my parents were illiterate, I could not find any written record specifying the exact day and year of my birth. By looking at historical events that occurred at that time of my birth, my parents estimated that I was born in August of 1941. August 23, 1941, is the date I gave as my date of birth when I registered for school.

Like all other children living in rural areas, I grew up playing water games with my friends, rolling around in the mud, wrestling, bareback horse-riding, and swimming in the river.

My parents were relatively wealthy, and they made every effort to ensure I grew up strong and healthy, meaning that they fed me all the milk and butter I could stand.

During the Italian occupation of Ethiopia, the invaders' mission was not simply a territorial expansion. They also wanted to spread their culture and religion. The Italians built a Catholic church in Emdibre town, Cheha district, and made a real effort to draw the local population into their fold. They also touted the benefits of modern education and built a school right by the church. Soon, many of the children in the area started attending this school.

After Fascist Italy was defeated, the newly re-established Ethiopian government took over the schools started by the Italians, opening them up to children of all faiths. However, many Orthodox Ethiopians resisted this modern approach to education and viewed these schools with suspicion, thinking that they were religious institutions primarily designed to convert their children to Catholicism. This general suspicion was encouraged and enforced by the clerical authorities. Most Orthodox parents generally opted to send their children to the traditional religious schools.

My parents were inclined to view the modern schools favorably, but they dared not go against the priests of the Orthodox Church since they were Orthodox. When it came time for me to start school, I did not go to the new schools with my friends. And, because there was not a traditional religious school near where we lived, I was sent to my aunt's place in Aklil, a place that the Gurage people call "*Ohiye*," between Eja and Muher. I did not start school until my second week there. That gave me time to become accustomed to the area.

The school was a large tree under which the pupils sat on flat stones. When I got there with my older cousin, we found the teacher seated on a three-legged wooden stool he'd placed on a leather mat, directly beneath the tree. He was murmuring the Psalms of David to himself. The pupils were huddled together around the teacher. His assistant stood pointing to the Amharic alphabet and a switch in the other. He would call out the letters of the alphabet in a loud voice, beginning with "*Ha*". The students would repeat after him.

"*Ha!*"

Then on to the next letter in vertical order, "*Le!*"

"*Le,*" the students repeated. My cousin handed me over to the teacher and left.

I spent the day shouting back the letters of the alphabet with the other children. When the teacher dismissed us, I went home. The next day would be writing practice. I was instructed to bring a broad leaf of the false banana plant, called *Koba*, that had been crisped with fire, as well as the writing tool I would use to write on the *Koba* leaf. The writing tool was actually a large thorn from the Agam plant which had to be blunted with fire, so it wouldn't puncture the writing surface. I told my aunt what my teacher had asked us to bring.

The next morning, I had the *Koba* leaf rolled up, and the thorn hanging around my neck on a string so I wouldn't lose it. When I got to school, the teacher scribbled some incomprehensible ciphers on the leaf, then handed it back to me saying I should closely copy what he had written. I did what he asked. In this manner, the next few months spent at the religious school went on in the same way.

The abundant food I was accustomed to was not available at my aunt's house. Food was a constant problem, and I was under-nourished. I began to lose weight rapidly, and my body became covered in sores. I grew steadily weak and soon reached a point where I had difficulty standing up. My childish enthusiasm disappeared. Gone was the loud running about, laughing and playing. Now all I wanted to do was hide and cry. School was not going well either. I was not having much luck with the alphabet counting. After six months of lessons, I could only parrot back "*Ha*" but could not identify it if asked.

My aunt and her son, seeing my condition, decided that it would be best to take me back to my parents. My cousin took me back to my parents on the back of a mule. When we got back home, my mother was alone at home.

My mother began to cry uncontrollably when she saw my terrible state. I was sick, emaciated and covered in sores. Hearing her wailing, our neighbors and relatives gathered quickly. Seeing the both of us, they too started crying. Overcome by it all, I cried as well.

In my father's absence, my mother swore that her son would never again go to any school, except over her dead body. She washed me with warm water and covered me in butter.

Later that evening after the cows were milked, I got some fresh milk and some butter. For a whole week, I was not allowed to leave the house as my body repaired and healed itself on the fatty diet.

My playmates and buddies who had been sent to the modern schools came to visit me. As we talked about our schooling, it quickly became evident that they were educationally far ahead of me. They knew both the Amharic and English alphabets and could also count to one hundred. I could not even distinguish the first few letters of the Amharic alphabet. My father could see that I was not doing as well as the other children and felt terrible about his earlier decision regarding my education. He waited a few more days until I had recovered a bit more, and then without consulting with my mother, he enrolled me at the modern school in Gubre. My mother, respecting my father, did not say anything when she found out.

The Gubre school was about four kilometers from where we lived. All the kids in my village who attended the school would meet after breakfast and gather in a big group. We would head to school playing and talking amongst ourselves. The distance did not bother us.

The government schools differed from the traditional ones in many ways. We had desks and chairs for one thing. The school provided us with pencils, notebooks, and free textbooks. The government paid our teachers. With the help of my teachers and friends, I caught up very quickly. I soon learned the Amharic and English alphabets. My success pleased my father very much. I continued to attend the school in Gubre for three years.

The Gubre school only went up to the third grade. Grades four through eight were taught at Emdibre. To continue our education, we had to go there. However, Emdibre was about twenty-five kilometers from my parents' home. It was not possible to commute daily. For me to attend the school in Emdibre, it was necessary to live with relatives living near that town. We went to stay with them, taking our own food and other needs loaded on a donkey. We also took a milk cow.

However, though we took our cow, our relatives did not allow us to have any milk from it. Even whey was a rare treat. On top of that, they started requiring us to do various chores after school. These chores included cutting grass, chopping wood, and drawing water. The tasks were time-consuming and cut into the time I needed to do homework.

Going to school without doing homework resulted in corporal punishment. The flogging was very painful.

Those of us from Sise all faced the same problem. We gathered together and discussed our dilemma. We decided that we would pool the resources we brought weekly from our homes to rent a house together, and this way we could focus on our education.

We prepared the weekly food provided for us by our families in turns. Every night, we would gather around the kerosene lamp and use its light to do our reading and homework until we fell asleep. Then we would rise early, eat our breakfast, and go to school.

Every week on Friday, we would make the trip back to our parents' homes and spend the weekend with them. On Sunday afternoon, we would load up our week's supplies and return to Emdibre. Occasionally, we would have to stay through Sunday and leave very early Monday morning. On those occasions, we would cover the distance at a fast trot and without rest. It was hard, but we would manage to get there in time for school. This became our routine as we progressed through grades four to eight.

Map 1. Born in Sise on August 23rd ,1941, the star represents the birthplace of the author and the circles represent the schools he attended.

CHAPTER 2
New Life

In 1956, my friends and I took the Ministry of Education's eighth-grade national exam and passed it. My friends and I went to the Ministry of Education to find out to which schools we had been assigned. My friends were all to attend either the Teferi Mekonnen, Wingate, or the Kotebe Haile Selassie secondary schools. I, however, was assigned to the Ambo Agricultural School. I did not even know where Ambo was and I certainly had no idea how to get there. I went to the central bus station in Addis to find out the cost for a ticket to Ambo, the departure time for the bus, as well as the travel time. Then, with that information in hand, I went back to where I would spend the night.

The next morning, I boarded the bus to Ambo. On the bus, I met and became acquainted with two students from other regions who, like myself, were heading to Ambo Agricultural School. We arrived in Ambo at 11 a.m.

The school was about two kilometers west of the town. We made the trip to the school in horse-drawn carriages. The carriage driver dropped us off at the entrance to the school, telling the guard at the gate that we were assigned to the school. The guard took us straight to the principal's office and left us in the principal's care.

The school principal assigned someone to help us get acquainted with the school and its grounds. Our guide first took us to our rooms, where we put up our luggage. Next, we went to the dining hall because it was, by now, lunchtime. Later that afternoon, our guide showed us around the campus. He then went over the school's rules and regulations with us and explained our rights and responsibilities as students. From the school's supply warehouse, we signed for and received our bedding, work clothes, and school uniforms. From the bookstore, we received our textbooks and stationery. Our guide then gave us detailed instructions on how to make our beds, water and electricity use, and time schedules for studying, meals, and relaxation.

Four individuals shared each dorm room, sleeping on two double-decker metal beds. Breakfast, lunch, and dinner were served on a regular schedule. The students from Addis complained about the food, but I loved it. Everything about the school, the buildings, student life, our classes, the beds, the electricity, even the weather, all seemed like a dream. Compared to my earlier life, it was terrific.

There were students from many provinces, but the greatest number by far were from Addis. Most of the Addis students were children of government authorities. Their comportment and speech mannerisms were different. They dressed differently as well, often wearing suits and ties. They changed clothes every day and sometimes even wore two shirts in one day. Some of them had two and three suitcases full of clothes.

In my region, I was considered the son of a rich man. But, all I brought with me was one suit, two khaki jackets and one pair of pants. I had two pairs of shoes, one leather and one canvas.

Some of my roommates from Addis, seeing my relative poverty, were not at all pleased to have me room with them. It was as though they feared that my country origins and poverty were contagious. They tried to have me moved to another room but were unsuccessful. In class, unless they had no choice, they avoided those of us who were from the countryside. In the dining hall, they sought out other students from the city, and would only sit with them. Seeing their disdain for us, we avoided them as well.

On Friday afternoons, half of the Addis students would have cars sent to take them home for the weekend. Those who did not go to Addis would often go to Ambo to get a break from school. Most of us from the provinces spent our weekends and other time off by going to the woods to study. Upon returning on Monday, the Addis kids would always make fun of us by asking, "Did you go to Addis? Did you go to Guder?"

However, they soon came to see that we were just as good as they were when it came to our class performance and overall results. We often placed at or near the top. Some of the academically weaker city boys who had shunned and mocked us now wanted our help and friendship. The city boys came to see that our origins and circumstances were not impediments. We spent the remainder of our high school years cooperating with and helping one another. It was a nice change.

At the end of 1960, I graduated from Ambo Agricultural School and passed the entrance exam for Alemaya Agricultural College. I passed both with flying colors.

I received a letter from the Dean of Alemaya Agricultural College, informing me that I had been accepted for the 1961 academic year. To reserve my place, around mid-August, I would have to go to the Ministry of Agriculture in Addis and confirm, in person, that I would attend.

When the school closed for the rainy season and having no place to stay in Addis, I went back home as I had every year. I found my dad desperately ill. He and my mother had divorced, and she had remarried. At home, there was no one but my little sister to care for my father. He was very ill and could not even sit up. He needed help to eat and drink. For three months, my sister and I cared for my father. Occupied with caring for my father, I missed the appointed date for the verbal confirmation at Addis in August. Finally, towards the end of September, my father was well enough for me to leave him in the care of neighbors and my sister. I hugged my father, and we both cried as I departed for Addis.

As soon as I arrived in Addis, I went to see Worku Mekasha, Director of Education at the Ministry of Agriculture. He told me he could not send me to Alemaya College. I had missed my chance. The foreign aid money that covered the position they offered me, by then, had been given to another student, and furthermore, classes had already started. I was told that if I wanted, I would receive priority for a position for the following year. I decided that I would be better off attending another college rather than wasting a year. After receiving a letter explaining my situation, I began my search. I made the rounds in Addis looking for a college to attend. But I was too late, and there were no available spots. Every college I approached gave me the same discouraging response, deadlines were passed, and positions were already filled for the current year. "Try again next year", they said. My chances for attending a college that year was gone.

I finally decided I would be better off going back to my village to care for my father for that year and try again for Alemaya next year. The very same day that I reached this conclusion, I ran into an elementary school friend on the street. After we exchanged greetings, he asked me what I was doing in Addis. I explained my situation to him.

He thought for a bit, and suggested, "Instead of going back home, why don't you enroll at the Army College? I think it would suit you just fine."

He told me that he had been selected to attend the college, but did not fancy becoming a soldier, and so had rejected the offer. Now he was hiding in Addis, where he taught at a private school. For my part, I had never even heard of the Army College. I had no concept of what it meant to be a soldier. I asked my friend to tell me everything he knew about the Army College.

"The Army College opened in Harar in 1957. It took three years to complete the course. Graduates would receive a degree and the rank of Lieutenant. They would also receive free government housing and a government car for their use. A graduate's salary would be equal to that of a university graduate." He quickly added: "I am pretty sure that they have not yet finished registration. You should head to the Ministry of Defense and ask for Major Addis Alem."

I liked his idea and did as he suggested. I headed to the Ministry of Defense. At the gate, I told the guards that I was a student and that I wanted to talk to Major Addis Alem. The head of the guards called one of his subordinates and told him to take me to Major Alem's office. I followed the guard as he led the way. He pointed at the door, then left me there and went back to his post. I knocked before entering. Once in the room, I bowed to Major and stood there waiting respectfully. The man I had come to see was impeccably dressed in a well-pressed khaki coat with shining golden buttons and epaulets. He sat in a large chair in which he swiveled from side to side. The Major asked me what I wanted, and I told him I wanted to enter the Army College. He asked me if I had my academic documents and transcripts. When I told him, yes, he asked to look at them. After looking through my papers, he looked up at me and asked why I was not going to Alemaya College. I explained my circumstances to him.

"Ok then," he replied, "there is a train that leaves tonight. Can you go?" I answered yes. He told me to be at the Legehar train station, with my luggage, at 6 p.m. When I asked for my papers back, he responded that he would be sending them on and that I would find them at the Army College. I thanked him and left.

When I got to the railway station, I found a group of students recruited from various schools. Standing in the middle of them all were two straight-backed, well-built officers. They were both impeccably dressed, wearing gray berets, and decorated as Major Addis Alem had been. One of them, seeing that I was standing apart from the students, detached himself and approached me. He asked me my name and school, which I told him. He looked in his notebook, then told me to join the others, and to listen carefully to the travel instructions that would soon be issued. I did as instructed. Later that evening the train arrived. We boarded and headed to Dire Dawa. We traveled all night and arrived at Dire Dawa at 8 a.m. Once we got off the train, we ate our breakfast at a hotel and we continued our journey to Harar using the Army College's bus. We reached Harar at noon.

CHAPTER 3
Cadetship

The Haile Selassie 1st Harar Military Academy was established in 1957 and staffed with elite officers and instructors from India. The academy's curriculum was on par with the best military academies in America, England, and France. The Emperor issued a mandate allowing the academy to take its pick of the top matriculated high school graduates throughout the country. This mandate provided the academy with the ability to select top quality candidates that might otherwise attend Addis Ababa University or other colleges in the country.

Students who left the military academy for health reasons were required to obtain documents certifying a legitimate discharge, and that they were unfit for military duty. The government issued an order stating that those who left without proper authorization would be barred from employment at any government or private organization in the country. They would not be allowed to travel to study abroad, and they would be tracked down by authorities and returned to the academy.

The first three classes of students, each one year apart, completed their three-year military education without any trouble. However, when the first class neared completion of their final term and were approaching graduation, the academy's commander issued the following orders:

- Graduates would receive a diploma and the rank of Second Lieutenant.
- They would receive a monthly salary of 120 ETB (Ethiopian Birr) as a Second Lieutenant, and an additional 90 ETB as a housing expense stipend.

The graduating officers were unhappy with this and tried to object to the new order. They were immediately issued an ultimatum; the failure to immediately comply with the new orders would be viewed as rebellion and acted upon accordingly. In the military, this could mean harsh penalties, up to and including death. Not having any options, the graduates accepted the new orders and graduated with the rank of Second Lieutenant.

However, this was not the end of the controversy. The next incoming class of students joined with the second and third-year students in an appeal attempt, again requesting that the original agreement under which they had enrolled be honored. They drafted a letter and sent it to the Emperor's top military commander. In the letter, the students stated their resolve to stop their education and not to attend any classes until they received a satisfactory answer to their appeal. During an assembly held to address the issue, the academy's commander ordered us to continue our classes, stating that he would look into the matter.

The students, including myself, stood our ground, remaining silent. The commander left the assembly hall in anger. Going to his office, he made calls to both the Ministry of Defense and the Emperor and told them of the situation. The Commander requested an investigative team be sent to Harar to look into things. A short time later, a team led by Lieutenant General Kebede Gebre arrived in Harar.

The Academy Commander removed us from our living quarters and transported us to Amaresa, about five kilometers from the College. They held us in a barbed-wire enclosed compound under armed guards and treated us like prisoners-of-war. One-by-one, we were summoned to go before the investigation team and state our case.

After a five-day investigation, Cadets Berhanu Bayeh, Ayalew Woldemeskel, Teferi Teklehaimanot, and Mersha Admasu were identified as the ringleaders of the rebellion and were condemned to go and serve in the Ogaden as privates. Cadets Goshu Wolde and Mamo Dinka were also reduced in rank to private. The rest of us received a severe warning and then were sent back to class.

There were rumors that the Academy Commander wanted all of us flogged and then dismissed from the Academy. It was also rumored that we were spared this punishment because of Ras Imru's intercession on our behalf.

The three officer-candidates sent to the Ogaden were granted a reprieve after a year and allowed to rejoin us. They graduated with their peers.

The Harar Military Academy was a strict institution. Trainees conducted themselves in strict accordance with military standards and the academy's curriculum. We trained hard day and night, in every terrain. Rifles on our shoulders, helmets on our heads, bayonets at our sides, our rucksacks on our backs, we would march and run, covering mile after mile. The road marches, endless drills, hand-to-hand combat, and bayonet training, pushed us to our limits. The academic curriculum was equally demanding. The workload was tough. They tested us every day. We talked, marched, dressed, ate and saluted like soldiers. All activity was strictly regimented. Punishment for even the slightest infraction was swift and brutal. Military Police would materialize instantly and drag the offender off to the brig, and sometimes to hard labor. Military life is very different from civilian life, and I had a hard time adjusting. On several occasions, I thought about leaving. Every time I went to Addis on leave, I vowed not to return. The only thing that kept me going back was the fact that I had no one to stay with while looking for another job.

I completed my three years of training, and in 1963, I graduated from the academy as a Second Lieutenant. I had still not come to realize, that I was meant to be a soldier.

After I graduated from the academy on October 13, 1963, I received orders to attend the Airborne and Commando Training School in Debre Zeyit. In May of 1964, I completed this additional training and reported to the 25th Battalion of the 3rd Division in the Ogaden, in a place called Degehabur.

Graduating cadets passing out parade, October 13th, 1963. Author:Front row, 1st Left

Pictured with Emperor Haile Selassie I on Graduation Day, April 13th, 1963
Author: Last row, 5th from Right

In preparation for a parachute jump

After the 1st jump, together with the trainers, Debre Zeyit Jumping District. 1964.
Author: 2nd row, 5th from Right.

At the time, Ethiopia was at war with Somalia in the disputed Ogaden. But, before I go on with my story, I want to give an overview of the land dispute between Ethiopia and Somalia.

There had been a long history of border disputes in the Ogaden region. An agreement between Ethiopia and Southern Somaliland had been reached in 1908, but there were still many unresolved issues. Continued discussions between Ethiopia and Southern Somaliland's colonial ruler, Italy, proved to be fruitless. Despite repeated negotiation attempts by the Ethiopian government, continued delaying games by the Italian government prevented any agreement from being reached. Therefore, Ethiopia's government took the matter before all the relevant world organizations of which it was a member and received legal and public recognition of its rights and sovereignty over the region. Ethiopia successfully reached an agreement with British-ruled Northern Somaliland in 1934.

In World War II, Italy, under the leadership of Mussolini, invaded Ethiopia. Then, in 1941, Ethiopian patriots with the help of the English and guerrilla fighters ended Fascist Italy's occupation of Ethiopia. When Italy finally left its colony in Southern Somaliland, the border disagreements between Ethiopia and Southern Somaliland were resolved, but the border markings in certain areas had not yet been put in place.

The English stayed on in Ethiopia for the next few years in an advisory capacity. The English argued that the French military presence in Djibouti was a threat to Ethiopia and that they should monitor the situation closely. The English government asked for permission from the League of Nations to occupy the Ogaden and maintain military oversight in the region. In 1942, they received that permission.

The English government then received permission from the League of Nations to administer Northern and Southern Somaliland. Lord Rennell was appointed as the new administrator for the region. He strongly advocated unifying the Ogaden, Djibouti, and North and South Somaliland into "Greater Somalia." Rennell had not made much progress toward achieving this goal when Ernest Bevin replaced him. Bevin was Britain's Foreign Minister and was a member of the commission set up by the victorious allies (the U.S., Britain, France and the U.S.S.R.) to determine the fate of Italy's colonies. Bevin advocated for Lord Rennell's plan. He argued that there were no discernible natural or legal boundaries between the Ogaden, Djibouti, and North and South Somaliland and that the existing borders were purely arbitrary. Bevin felt that Greater Somaliland should be created using more natural and realistic boundaries.

There was one thing Bevin did not understand. Before the colonial Europeans' arrival in the Horn of Africa, the Ethiopian and Somali people were one people under one government. He obviously did not bother to learn the history of the region. It was the Europeans who had created the arbitrary boundaries he so disliked.

Between 1941 and 1948, the Ogaden remained part of the colonial entity first established by the Italians and then continued by the British. Then in 1948, under an agreement with the Ethiopian government, Britain returned the Ogaden to the Ethiopian government, which resumed administrative control from Addis Ababa.

The English, on the other hand, when they left Northern Somaliland, had already sown the seeds of future troubles, thus helping to ensure the future instability of the entire region. The English formed the Somali Youth League in 1943 (possibly the first Somali political party), the Somalia League, and the Greater Somaliland Party. All of these entities had, at their core, the goal of creating a Greater Somalia. Forming these organizations guaranteed future violence with Somalia's neighbors. The purpose of all these groups was to unify all Somali speaking, ethnic-Somali, and Somalis who were followers of Islam under one government. The British-proposed Greater Somalia concept lit a fuse which would, in time, cause the whole region to explode in violence. It would begin the implementation of Bevin's masterplan to annex territories from Ethiopia and Kenya and to absorb them, along with all of Djibouti, into Somalia.

One of the criteria for Somalia's acceptance into the U.N., when it achieved independence in 1960, was agreeing to the resolution of the border disputes with Ethiopia, that the international bodies had already ratified.

Most of the newly formed independent African nations

established their national boundaries in this manner. When the former Organization of African Unity (OAU), now the AU(African Union) formed in 1963, both its charter and the U.N. charter, stated that the newly independent countries would maintain the borders they had set at the time of their independence, or that had been established by the colonizing countries. However, the Somalians declared that they were not subject to the international treaties and laws and that they would resolve the matter on the battlefield. Somalia declared war on Ethiopia in 1956.

Somalia launched a military offensive through Tegochale and Warder with the goal of placing Jijiga under their control. At Tegochale, the Somalians overran the small contingent of border police and started towards Jijiga. The Third Infantry Division, compared to the enemy force, was numerically inferior. It was, however, a disciplined, well-trained, and patriotic division. Upon learning that the Somalis had overrun the border guard unit, the 3rd Division moved immediately towards the main Somali force and halted its advance. The 3rd Division split its forces, with the first half maneuvering to stop the main Somali offensive and the second half circling in a flanking movement to push the Somalis back.

Once it gained the initiative, the 3rd Division began to pursue the fleeing Somalis. It had Hargeisa in its sights. Those who participated in the action stated that had they continued the pursuit, there was no question that Hargeisa would have come under Ethiopian control. However, as the 3rd Division was routing the Somalis and closing on Hargeisa, Emperor Haile Selassie issued an order that it immediately halt the advance and withdraw to within Ethiopia borders.

The Somalian government, having been so easily defeated on the battlefield, now turned to guerrilla warfare, and began sending agents into the Ogaden and the Bale Province to sow the seeds of ethnic and religious discord among the civilian population in the region. It welcomed supporters into Somalia and trained them in guerrilla warfare, supplying them with arms. These insurgents, now taking their orders from the Somalian government began attacking isolated settlements, defenseless civilians, and convoys in the Ogaden, Bale, and Sidama provinces. They blew up bridges and cut down electric poles and lines. They routinely opened fire on and set off explosives in towns, spreading terror throughout the population. All infrastructure and populated areas now needed military protection.

The responsibility for protecting the country south of the Awash River, the whole of Harerge province and the Ogaden, virtually everything from Gode to Djibouti, belonged to the 3rd Division. The territory it was charged with protecting was just too vast for one division. Therefore, available military protection for any one place at any one time was significantly limited.

I joined the 3rd Division after its victorious campaign against the Somalis when insurgent and guerrilla forces had become widely prevalent and active throughout the Ogaden. Not long after I arrived there, I was assigned as a platoon leader. There was a standing order from the 3rd Division Commander that every battalion was to keep strict control over the area they had responsibility for by sending out regular patrols to foil guerrilla attacks.

I had barely had time to acquaint myself with the men under my command when it became my turn to lead my platoon out on patrol.

A platoon on patrol must be hyper-alert and vigilant. It must be entirely and continuously prepared to engage with the enemy. The 25th Battalion had not yet had a single encounter despite many patrols. Therefore, its commanders, lulled into thinking that the area was safe, had not maintained their men's combat readiness. The entire battalion was not well led. The platoons sent out on patrols would merely go through the motions and did not bother to follow proper combat protocols. This laxity became reinforced over many such patrols. The men would head out on patrols and return without any sense of urgency or military professionalism.

As a fresh graduate of Harar Military Academy and the Airborne and Commando Training School, I was unaware of the lax state of the battalion's combat readiness. I merely prepared my men as I learned to do at the academy before we went out on patrol. Each squad leader understood his responsibility and what would be expected of him should we engage with the enemy while on patrol.

To prevent the entire platoon from coming under surprise attack, I sent a small squad of scouts to take the point. They stayed between fifty and one hundred meters in front of the platoon to report on any signs or enemy activity or to delay a sudden enemy assault. My men knew to stay alert and scan the horizon, to keep their eyes and ears open and their weapons at the ready. They understood that the best way to survive was to see the enemy first and close with him before he could engage us. Prepared as best we could be in the time we had; I led my platoon out into the unknown thorn-and-thistle infested land that was still very new to me.

I was a real novice. But, this was nothing new to the men of the 25th Battalion. The 25th Battalion was a battle-hardened fighting force. They had seen the action both at home and abroad, most recently in the Congo. They were not exactly thrilled at having to take orders from a green lieutenant, and only grudgingly complied with my commands. Still, our patrol proceeded professionally, and we arrived at our assigned objective at about noon.

After some reconnoitering to make sure that we were alone, I assigned one squad guard duty and ordered the rest to have lunch. When the first two squads finished eating their rations, the squad that was on guard duty was replaced by another so that its soldiers could eat their meal. After lunch, we rested a few minutes more, then continued on with the patrol. The heat was overwhelming, and on the way back, the men began exhibiting signs of its fatiguing effects. It was clear that my men's current conduct would make us vulnerable. I summoned the Platoon Sergeant and after discussing my concerns with him, sent him back to ensure that the squad leaders regained positive control over the men.

Just as he moved off to do this, the scouts on point started to come under enemy fire. This firefight was the first time I had experienced hostile enemy fire. It was quite unnerving when a round went whizzing past my ear. I can indeed say that any flesh-and-blood being not instantly animated by the experience of bullets hurtling toward him must already be more dead than alive. No amount of training can prepare you for the shock of real combat. The realization that the bullets flying toward you are meant to kill you is extremely disconcerting. The only choice you have as a soldier is to settle your nerves and fight the battle with as much composure as you can muster. And, that is what I did. I quickly gained control over my fear, took stock of our situation, and began directing my men. We needed to move swiftly to a better position.

"3rd and 4th squads, you will remain where you are and provide cover fire on my order. 1st and 2nd squads, upon my command you will quickly counter-attack and occupy the high ground to our immediate right." Everyone had their orders.

"3rd and 4th squads, open fire! 1st and 2nd squads move forward!" I had ordered the Platoon Sergeant to stay with the 3rd and 4th squads while I led the 1st and 2nd squads in the assault. We started our counter-attack. We quickly seized control of the high ground to our right, and from there we were able to control our enemy's movement.

I suspected that the enemy might have laid a trap and planted mines. They were probably expecting my men to move straight towards them. My 3rd and 4th squads could get cut to pieces in a minefield. Therefore, I ordered the 1st and 2nd squads to begin providing cover fire and directed the 3rd and 4th squads to move to the left to occupy another section of elevated ground. Once that high ground was secured, I had the 1st and 2nd squads launch flanking attacks, cutting off the enemy. The Somalis were forced to abandon their position and were soon in full retreat, scattering in every direction. We pursued them for a bit. But again, I suspected that they might be luring us into an ambush kill zone. It was definitely a place we did not want to find ourselves. Therefore, I terminated the pursuit, and we returned to complete our patrol and return to base.

As we approached the base, my troops could hardly control their happiness. They began singing Ethiopian war songs, *fukerana shilela*, as we grew closer to the battalion encampment. It was with a great sense of pride that I ended my first day on patrol and my first taste of battle.

Our patrols continued for some time and through various areas. My platoon participated in assaults, often with the vanguard. Occasionally, we served as a reserve force or on standby. For the duration of our time, my platoon performed their duties professionally and with honor.

I turned my attention to making good use of our downtime when we were not out on patrol. In close collaboration with the battalion operations and combat-training officers, I devised and implemented programs to keep my platoon in the best fighting condition. I worked hard to keep their morale, combat proficiency, and discipline at the highest levels possible. In our spare time, we cut wood and built shelters daubed with mud. This work provided my men with more comfortable accommodations than just the open field where they had previously eaten and slept. I spent the next six months in this manner, solidifying my relationship with the men, as well as my command.

My 1st mission with the 25th Battalion after our return from a combat patrol in Ogaden on August 30th, 1964. Author: Front row, Right.

CHAPTER 4
Special Forces Training

I SPENT SIX MONTHS at Degehabur. One day, while conducting routine training exercises, I spotted a small plane that had come from the direction of Harar. Circling the landing field at Degehabur, the aircraft started into its landing approach. Planes were an unusual sight in this part of the Ogaden, and the soldiers watched its approach with fascination. I stopped our training so that the men could indulge their curiosity. I did the same, intensely interested in the way in which the plane touched down during landing.

After the plane landed and cut its engines, we resumed our training. Thirty minutes after the plane landed, a Military Policeman approached me and saluted. The policeman informed me that the commander requested my presence in his office. I instructed the officers to continue the training and followed the MP back to the camp. I knocked, entered the commander's office, and saluted. The commander was talking to the pilot and co-pilot of the plane. It turned out that I knew the pilot. After we greeted each other, Colonel Belete Haile, the commander, addressed me.

"Lieutenant Tesfaye!"

"Yes, sir."

"Our superiors have chosen you to work with the Israelis to set up a new unit. You will leave immediately with these men for Dire Dawa. Quickly gather up your belongings and return."

I saluted and left his office. As I rushed to my quarters, I began thinking about this exciting turn of events. I was stunned by what the Colonel just said. It seemed too good to be true.

I must be dreaming, I thought. In my wildest dreams, I never would have believed that I'd leave Degehabur so quickly and with such an important assignment.

I decided that such matters are best left in the hands of God. Quickly, I grabbed whatever clothes and other personal items I could find in my quarters, threw them into my suitcase and rushed back to the commander's office. We took his car to the airport. When I got out of the vehicle, the Colonel embraced me and wished me good luck as we parted company. Trembling with excitement, I boarded the plane. I still found it hard to believe I was leaving. The plane's door was secured, and the engine started. The aircraft accelerated down the runway and lifted off. When the plane finally lifted clear off the ground, reality finally dawned on me. I was leaving Degehabur. And, as the aircraft climbed into the air, I thanked God for allowing me to leave the Ogaden safely.

It was a one hour flight to Dire Dawa. After we landed, the plane taxied towards the terminal and stopped. The pilot cut the engines, and descent ladders were brought up. I climbed down. Inside the terminal, I was met by a group of Israeli and Ethiopian officers who were waiting for me. They took me to the encampment.

Dire Dawa was like another world. Electric lights rather than kerosene lamps, spring beds with cotton mattresses instead of canvas cots, hot water and bathtubs replaced the bucket and tin-ladle baths in the field. And, the food…proper *injera* and *wot* prepared by a cook replaced the constant diet of inedible crackers …softened by crushing them with our helmets and soaking them in water or tea. Best of all, now we were coming and going without the constant fear of attack. Compared to the hell of Degehabur, this was paradise.

Three days after my arrival in Dire Dawa, training began. It was a small group consisting of eight officers and thirty non-commissioned officers (NCOs). Four of our instructors were Israelis.

The training consisted of twenty-five percent classroom theory and seventy-five percent practical training. We used live ammunition and explosives. Injury and death were a real risk and accepted as part of the training program. The hard work never stopped. If anyone was injured or killed, trained field medics would move in and tend to the wounded. They would also transport the dead away and prepare the bodies for burial. There could be no interruption in training. A five percent injury and casualty rate were expected, and the trainers would not be held liable. Eighteen-hour workdays were routine.

Our Israeli trainers accepted no errors. The smallest of mistakes received strict attention and scrutiny. We learned that tolerating the smallest failings now could lead to much more significant losses later when real lives were on the line. Everything we did in training was as close to the real experience in battle as it could be. We conducted every training session as though we were in an actual fight, and as though the real enemy was right there, trying to kill us. We were not allowed the mental relaxation that could occur from the knowledge that this was just simulation; that it was not real. During training, we were always trained to be alert and prepared as though we were in battle. Our gear and ammunition were kept with us and complete. Our water bottles were full and strapped to our sides. Our helmets stayed on our heads. Also, we always carried burlap bags filled with sand in our packs to approximate the weight of the combat load we would have to carry on our persons during the actual battle. This regimen applied to all trainees, officers as well as NCOs. We were all outfitted and weighted down in the same way, all the time.

Our training was thorough and robust and covered a lot of subjects. Just a few examples are:

- Explosives training.
- Rappelling.
- Exiting and entering vehicles at high speeds.

- Combat swimming.
- Scaling buildings.
- Negotiating obstacles of all types.
- Hand-to-hand combat.
- Survival techniques.

I will elaborate a bit on the first and last skills in the above list.

The use of explosives plays a crucial role in many special operations missions. A special forces soldier must be very proficient in the use of explosives. For example, he must know how to calculate the type, amount and proper placement of explosive power necessary to destroy structures such as bridges, bunkers, or dams constructed of concrete or steel. One of the special forces soldier's primary tasks is to find the most vulnerable parts of any installation, any target, and destroy or disable it. These targets can include electric poles; railroad tracks; factories and power generators. In short, any military target that can be damaged or destroyed by explosive and incendiary devices are his targets. His training must give him the ability to distinguish and understand the chemical, structural and technical properties of his targets, as well as the destructive potentials of various explosives, and how to employ them accurately on his own and without any one's assistance.

Survival training is at the core of any special forces soldier's skill-set. It is a central pillar of the special forces training regimen. The essence of survival skill involves knowing where you are and knowing how to find and use whatever the environment in which you find yourself provides to help you survive. Any situation can provide you with the essential elements of food and water if you know where and how to look. There is always some living thing, no matter where you are, that can help you increase the odds of your survival.

For example, of the three thousand or so legless and reptilian creatures found on land and in water, ninety percent are not poisonous and can be eaten. You can extract some form of food or water from all kinds of plants. The knowledge of these survival possibilities and the skill to exploit them are drilled into a special forces soldier during training.

The soldier is trained to have the knowledge, skills, and the pragmatic disposition which will allow him to survive by shedding the various cultural and personal prejudices that make it difficult to do whatever it takes to survive. The experienced soldiers and trainers would often demonstrate this mindset for us. Once a month, they would capture and eat various reptilian species, birds, and four-legged mammals of all kinds, including monkeys. They would have us do the same. It was clear that a formidable mental barrier had to be overcome, especially given that Ethiopians have strict codes when it comes to edible food. But survival and the mission must take precedence over propriety in combat. We all complied with our training eventually and learned to do whatever it took to survive and complete the task.

We had less trouble with plants. The only thing was distinguishing those that were edible from those that were poisonous.

As a final test after completing survival training, the special forces soldier would be ordered to cross the three hundred kilometers of the Adal Desert and return. Then the trainee would be assigned to infiltrate the base and simulate the sabotage of a designated target on the base. The trainee then had to make a clean escape. He is given no food or water. All he has is the order, and it was up to him to use his survival skills to complete the mission.

This test is extremely brutal and pushes the trainee to the edge and beyond. Success requires endurance, patience, and courage. Both parts of this exercise were brutally tough and dangerous. Let me share with you what might typically transpire during this test.

The three hundred kilometers of the Afar Desert is harsh and uneven terrain with hills, gullies, and ravines. It is covered with various low-lying desert flora. The hellish heat starts at dawn. Without any food or water, the trainee must go into this desert, trusting this inhospitable nature to provide his needs. He is not allowed to carry any weapons. He will have to depend on his trained body and mind alone. To complete this trial in the allotted time and return to his unit, the trainee must have a single-minded determination to reach his goal. He must be in top physical condition, and he must have iron discipline. He will need every bit of the training he has received to survive.

He must avoid sweating caused by the hot temperature. Having no water to replace the sweat, the soldier must avoid movement during the hottest parts of the day and only move at night. Only when the sun starts going down, and the air begins to cool can he safely move.

Trainees are assigned different zones of the desert. There would not be more than two in one zone. While crossing the desert, trainees had strict orders to stay away from vehicle roads, railroad tracks, and footpaths. For training purposes, we had to look upon the local people as enemy combatants. We had to keep ourselves hidden from their view. To avoid sprains or broken bones, we always had to be aware of the ground and where we placed our feet. Any such injury could have the gravest consequences. Walking in the desert meant heat, sweat, and friction which could cause the skin on our feet to blister, tear and get infected. It doesn't take too much imagination to understand how painful it would be to walk on sore, bleeding feet. It was imperative to take every precaution to keep that from happening.

All movement would stop at 6 a.m. Then came the search for water. Trainees would try to find a tree near a dry riverbed, hunker down in its shade and look for remnant water just beneath the surface. If that failed, the trainee would try to cool his feet and body in the wet sand. If there were no water, they would chew on the leaves of various plants to generate saliva and keep their throats moist. The succulent aloe plant could serve as both food and water, once the thorns are scraped off. The sound of bird calls, or the direction of their flights, was also a good indicator of nearby food and water. The sounds and movements of other animals, or the hoof marks of domesticated cattle, were also signs for which we trainees would be on the watch.

If all the above failed, the trainee would wait until dawn and try to squeeze what little moisture he could from his clothes into his parched throats. Trainees were strictly forbidden to drink urine under any circumstance.

When I took this test, we all completed the trial ahead of the deadline. Other than numerous scratches by desert plants, or lacerations to our feet, none of us had significant injuries. We all got a good measure of our physical hardiness, our mental toughness, and of our ability to exploit the environment to survive all the difficulties and complete the assigned mission.

The second part of the test was to penetrate an encampment, simulate the sabotage of selected targets, and then make a clean getaway. We conducted these exercises against armed guards with permission to kill. The training could easily result in us getting killed. These exercises were important because they helped us gain intelligence about the strength of our defenses and our capacity to detect and foil just such an attack by enemy insurgents.

The infiltration exercise would not commence until the permission of the officers in the responsible sector, as well as the commander, had been received. The officers involved would be asked not to heighten, or in any way modify, the security arrangements which they would typically employ before consenting to this infiltration exercise.

On the day of the exercise, only the commander involved and the officers in charge of Tactical Operations and Information would be informed that the activity had commenced. If the Special Forces trainees reached their target, the officers would ungrudgingly do whatever we asked them to do. They would also ensure our safe exit.

This exercise was first carried out at the 5th commando battalion military camp at Adele. On the day, the commander of the 5th commando battalion broke his word. He ordered the entire army to assume a heightened security posture and gave strict orders for the reinforcement of all the guards. We were now faced with a solid wall of armed men that a mouse couldn't have gotten past. We couldn't find a way through. Unfortunately, it looked like we would be forced to terminate the exercise. Defeat was not something we were willing to accept. So we improvised.

We assigned the main group of our Special Forces to assemble at a prearranged point and wait to give defensive protection if needed. A smaller diversionary group was sent to another location to draw defensive fire. The plan was to make it look like we had been detected while trying to penetrate that location. The men comprising our diversionary action started firing in an attempt to draw return fire. The guards there obliged and returned fire. Other guards along the defensive perimeter erupted in joyful anticipation of our detection and humiliation. Eager to see, the special forces fail, many of the men left their assigned positions along the cordon and rushed in the direction from which the gunfire exchange had sounded. They wanted to see who we were and revel in our capture. We quietly waited until they kindly removed the barrier they had erected and then slithered on our bellies past them, all the way to the commander's bedroom. The commander was not pleased. Once he ascertained who we were, he drove us out of the compound himself. Our exercise with the 5th commando battalion ended successfully and without any injuries.

It was then decided that the exercise should be repeated at the 10th mechanized brigade in Jijiga. Again, we informed the Brigade's commander that the primary objective of the exercise was not to test the fighting ability of his command, but rather the reliability of his unit's security posture. The commander agreed that the error of the 5th Commando battalion would not be repeated with him.

It turned out to be worse than the disaster at Adele. On the training day, he alerted the entire Brigade of the exercise and issued a strict order to his full command to preserve the honor of the Brigade. The command was issued with warnings of harsh consequences. The guard detail we were to probe had been reinforced five-fold. We tried to get close from several directions. There was no way.

We were again forced to create a diversion to distract the troops and create an opening. Again, we selected some of our group to go some distance from the point we had chosen to penetrate the compound and draw the guards' attention and create the illusion of detection and capture. This time, the deception was verbal. Our diversion team raised a terrific cry, as though they had been detected while attempting to break in and were in danger of their lives. They begged for mercy.

"We are your brothers; we are on the same side! Please, don't shoot, don't kill us. Please! You have won already. We are already in your hands, your prisoners, is that not enough? Why are you hitting us!" Our men would then vanish from the spot where they had been loudly crying and begging. The guards who heard the noise abandoned their posts and came running to see the captives. Like the cruel guards at Adele, they were in no mood for mercy.

"Lay on, beat them! They have asked for it. They can't possibly get enough!"

Once again, as they rushed past us to see the captives, we slithered past them. This time we went straight into the commander's office and smartly saluted him.

"We are here."

The commander admired our skill, but he was not pleased with his men. As he drove us out of the compound, a guard at the gate attempted to stop us.

"Stop! Stop! Reveal yourself!" the guard ordered. The guard's sudden initiative was met with a terrific hail of insults by the commander. We exited the compound safely.

These exercises had a two-fold outcome. We were able to demonstrate to our side the necessity of improving the security around the commanders in the vicinities of both Dire Dawa and Harar. For our part, we learned that a solid wall of men could not stop us. We could find a way between their legs. It led us to understand that no security was impregnable and that to complete our mission, we would find a way through.

The training and the exercises ended very satisfactorily. Now it was time for us to put our skills to good use and turn the table on our enemy, giving him a taste of his own medicine. We would infiltrate his territory and wage offensive and defensive battles of subterfuge and sabotage.

CHAPTER 5
Mission in Borama

IN EARLY 1967, Ethiopia's border battles with Somalia became increasingly more violent. The situation was very fluid, and tensions were ratcheting up quickly.

The Somalian Army established close working relations with the Western Somali Liberation Front. They sent trained agents into the provinces of Ogaden, Bale, and Sidamo to kill civilians and destroy infrastructure using mines and bombs. They attacked convoys and supply trucks delivering provisions to the Ethiopian Army as well as basic necessities for the people of the Ogaden. These agents would locate and attack isolated groups of civilians throughout the region. They blew up roads, railroad tracks, and bridges. Lines of communication were frequently disrupted. Their frequent ambushes and extensive use of explosives forced us to abandon the road from Negele to Dolo, and resupply our battalion at Dolo by plane.

They kidnapped and held the ethnic Somali elders and tribal leaders who had been appointed to administer the Ogaden region and confiscated the vehicles provided for their use by the Ethiopian government.

The Sudanese government soon aligned itself with the Somali

government and provided the Eritrean Liberation Front (ELF), or Jebha, an office in Khartoum from which they coordinate their efforts. The ELF set up a radio station and began the dissemination of anti-Ethiopian propaganda over the airways. They also began to recruit, train and arm ex-patriates and refugees, which would then be sent into Ethiopia to spread terror.

Soon, they were targeting the domestic and international flights of Ethiopian Airlines and hijacking flights and directing them to destinations of their choice, often harming the passengers. These attacks were very costly to Ethiopian Airlines and if unchecked, might soon lead to the failure of the airline.

The constant acts of terror and violence were quickly destabilizing the population. People could not move about in safety. Fear and uncertainty gripped the entire country. The civilian and police administrators eventually abandoned their posts and fled to safety in Dire Dawa and Harar.

The Ethiopian government was desperate to find a way to stop the increasingly violent attacks by the Somali government and its allies. One of the retaliatory plans considered was the kidnapping of Somali civilian administrators in Somali border settlements and towns. This plan was how Borama became targeted.

The Special Forces were established as a defensive fighting unit. For administrative ease, Special Forces had been attached to the 3rd infantry division. Once the high command decided how to proceed, they outlined their priorities and tasked the commander of the 3rd infantry division, Brigadier General Abebe Gemeda, to command the operation. Upon receiving the orders, General Abebe contacted the commander of the Special Forces, Major Kassa Gebremariam, and tasked him with immediately assembling a strong team and reporting to 3rd Infantry Division Headquarters in Harar by 9 p.m. that same day.

The Special Forces soldiers were all extremely qualified. Major Kassa handpicked six officers and eight NCOs who he thought would be best suited for the mission. I was one of these men. Within ten minutes of receiving the word, we were all in his office. Major Kassa proceeded to give us what information he could at that time.

"I do not have the details of your mission at this moment. As soon as we arrive in Harar, we will be briefed on the specifics. We have orders to report to the 3rd Infantry Division for an urgent mission. You are to go to your quarters, dress in casual civilian clothes and leave behind anything that can be used to identify you, such as identity documents, written documents, money, jewelry, or wedding rings. Then return here."

We soon returned dressed in casual pants, shirts, and light jackets or sweaters. Waiting to transport us were two large Land Rovers with extended cabs and extra-range (200 liters) fuel capacity. From the armory, we were each issued Uzi submachine guns with silencers, bayonets, two grenades, and two-meter lengths of rope. We climbed into the Land Rovers in two groups of seven: three officers and four NCOs.

We suspected that there might be spies following our movements, so we waited until it started getting dark before leaving the base and driving through Dire Dawa. The first car departed at 7 p.m. and took off in the direction of the French Hospital in front of the train station. Keeping the hospital to its left, it took side streets towards Dire Dawa Stadium. Once past the Stadium, as it neared the Prince Makonnen Secondary School, it got on the Dire Dawa-Harar asphalt road and headed to the rendezvous spot.

The second Land Rover left ten minutes later and headed towards the cotton factory, then through Gende-Dibo, and Kefira, and to the Stadium, keeping the Ras Hotel to its right. It would get onto the asphalt road right in front of the Stadium and exit town and make its way towards the rendezvous. As planned, both cars met at the link up point and then headed to Harar, arriving about 8:30 p.m.

Once we reached Harar, the 3rd Army Commander, General Abebe, led us into a small room to brief us on the mission. He began with a brief overview of the current political situation. Then he went on.

"Our government has made every effort to resolve things peacefully.

The Somalians are not responsive to our efforts. The only option we have now is to strike back and hit them hard. You men have been tasked with a mission that will do just that. You will go into Borama, where you will kidnap the regional administrator and bring him back alive. Our regional administrator in Jijiga has been briefed and will meet you there to provide any intelligence you require to succeed in your mission. He will provide road maps, translators, or other additional supplies that you might need. You will meet him at the base of Karamara Mountain, five kilometers west of Jijiga. If you encounter any difficulties, contact me immediately. I will be closely monitoring your mission. We will meet again soon after your successful mission."

We started for Jijiga right after eating the dinner provided for us and reached the base of Karamara about 1 a.m. As planned, we met the regional administrator for Jijiga, Balambaras Fekade and proceeded carefully to the resting place he had prepared for us.

We spent the next day in isolation, resting. No one came near us except for the cooks and attendants who saw to our needs. No one was allowed to leave the compound. That evening, as darkness fell, Balambaras Fekade, accompanied by a second man, showed up in his car. He brought clothing, shoes and other items that were typically worn or used by the people native to the area into which we were going.

Balambaras Fekade was an ethnic Amhara, but he had been born and raised in Jijiga and therefore, was thoroughly familiar with the local customs and was fluent in the language of the Somali people. He showed us how to wrap the sarong-like garments around our lower bodies, how to wield the sticks always carried by men, and the proper way to present ourselves. He did his best to help us look like people native to the region.

Our guide was an ethnic Somali and fluent in Amharic as well as Somali. He was familiar with Borama, and its regional administrator whom he could physically identify. While waiting for the streets of Jijiga to empty, we absorbed the necessary intelligence and finalized our preparations. At 11 p.m. we left Jijiga and headed for Teferi-Ber, arriving at our destination in an hour.

Teferi-Ber is a small border town. When arrived, the air itself seemed as dead as the villager. Neither villager nor tree stirred. There was an eerie quiet over everything. We hid our vehicles in some woods, then crept stealthily to the little *tukul* prepared for us. We slept. We spent the entire next day hidden there, gathering and studying intelligence, and finalizing our plan. We went over every detail including Borama's topography; the exact location of the administrator's house and his office and his schedule. We memorized any distinguishing marks on these buildings, the make, and color of his car. We needed to know about his chauffeur and bodyguards, how many he had and how they were armed. Whether his wife, children, and servants were in his house. And if so, how many they numbered? Was his home guarded? By civilians? Or, by the police? How many guards would there be and how were they armed? Did the administrator go straight home from his office? And if not, where did he spend his evenings? Who might he visit? What time did he usually get back to his home? Much of this intelligence we gathered from residents of Teferi-Ber, who often traveled to Borama, and would pass information in both directions.

Borama is about fifteen to twenty kilometers from Teferi-Ber. Dressed in our local garb sandals on our feet, sticks appropriately carried across our shoulders, we struck out for Borama at about 7 p.m. We did our best to pass as native inhabitants. Around 11 p.m., without having encountered a single soul, we reached Borama.

Approximately one kilometer north of Borama, we located some high ground covered with corn. Hidden in the corn near the foot of the rise, we had an excellent view of the town. Also, the corn covering the high ground continued right up to the edge of town. We were able to move right up close to the town to conduct our surveillance with little risk of being discovered.

We could see the comings and goings of the town's inhabitants. Borama was a small town with scattered buildings, so it was easy to spot people's movements. We could see a good number of large cargo trucks in motion. The only privately owned automobile was the regional administrator's short-cab Land Rover. To the west, we could see the military camp. While we knew that this was the region's protective police force, we could not determine its actual size. The regional administrator's house, a villa with its large compound, stood by itself to the east. It had a wall around its perimeter. His office was located in the center of town which stood out against the surrounding buildings. A steady stream of people moved in and out of this building. We could also see several groups of men standing in its compound.

The corn was ripe. So at dawn, the local farmers posted boys with slings, as their lookouts and these boys began to battle with the birds, their sarongs loaded with stones. Once the bird warriors assumed their positions, they started launching canon-ball sized rocks in every direction. The missiles began crashing around us. The stone slung by the strong youth sailed harmlessly over our heads. But the ones thrown by the weaker boys were landing right on us. We did not dare move. The young bird warriors would have detected any movement by us and cause them to raise the alarm. Our only option was to remain where we were and endure our stoning. Not one of us escaped getting dinged at least four or five times. Some even had bloody injuries. Fortunately for us, as the day lengthened our bombardment began to slow. Perhaps the bird-warriors strength was being sapped by the sun, or their supply of stones was dwindling. Very carefully, we took turns getting as close to the town as we could and staying well hidden in the corn. We continued to monitor the activity around the regional administrator's house and office as best we could. Around 9 a.m. that morning, we observed the short-cab Land Rover as it left the administrator's house and drove to the office where it was parked. We saw no other human activity at his residence.

We concluded that he lived alone and that his wife and children were probably in Hargeisa. We discovered that he dined out at a hotel and that he did not return to his house during the day. He ate his dinner in the evenings away from home and only returned to his house to go to bed. He also drove the Land Rover himself and did not have anyone guarding him. Daily, he was assigned two policemen armed with Kalashnikov automatic rifles to guard his house.

At noon, we saw the administrator drive the Land Rover, from where it had been earlier parked outside his office, into town. He parked outside a house where he often took his meal. At 3 p.m., he returned to his office. We now knew where to target him. All that was left was to devise the final plan to get it done.

We came up with three different options. The first was to grab the administrator as he was leaving the house where he took his dinner just as he was getting back into his car. The second was to wait until he reached his home, and grab him as he was about to enter. The third was to wait until he was in bed, then kidnap him from the room as he was sleeping. We weighed the pros and cons of all three. We eventually decided on the third option thinking it would be the best option for keeping any alarm or noise at a minimum. We worked out the final details and then split into three groups. Group one would neutralize the guard. The second group would stand watch, and the third group would take the regional administrator into custody. Each group clearly understood its role and responsibilities.

At 8 p.m., we approached the administrator's house in advance of his arrival. We'd learned where his guards were posted, and the strengths and weaknesses of their positions as well as the best way to gain entry. We also studied the habits of the neighbors and how to avoid any problems with them. Our escape route after kidnapping the administrator was laid out. We had even estimated the time it would take the police guards in the camp to reach the house, and the direction from which they would approach, should an exchange of gunfire alert them. The team tasked with capturing the administrator was in place. We were ready.

Our guide would stay with the kidnap team, awaiting the signal from us. As the administrator neared his house, the guide would approach him, pretending to be one of his relatives, calling him by name. He would tell the administrator that a tribe known to be an enemy tribe had attacked and robbed him. The goal was to get the administrator to invite him into his house. At that moment, we would signal, and the kidnapping team would move in quickly and subdue the administrator. We would tell the administrator that while in our custody he would be well-treated if he cooperated. If not, his suffering would be extreme. The nature of our hospitality depended solely on him.

The hour the administrator was expected to get home was approaching, and all our teams were in place. 9 p.m. came and went. Then it was 10 p.m. The regional administrator still had not shown.

By 11 p.m., we suspected that someone had tipped him off and that he had fled.

Prolonging our wait in enemy territory was too dangerous for us. We decided to exercise the next best option. This option included eliminating the guards. Therefore, quickly and silently, we killed both guards. Posting lookouts, we entered the house to take anything that would be helpful to us. We took pictures of his wife and children and various documents. Then we booby-trapped some doors and the bodies of the two dead men. At midnight, we headed back towards Teferi-Ber arriving there at 5 a.m.

There, we removed our skirts and put on our clothes. Immediately, we got into our Land Rovers and headed straight back to Harar. We reported to the 3rd Infantry Division Commander, General Abebe Gemeda, then returned to our base in Dire Dawa and spent the next night there.

The Somali government broadcast the story about how Ethiopian agents had infiltrated Borama and attempted to kill the regional administrator and the chief of police. They threatened vengeance immediately. However, after our attempt to kidnap the administrator, Somali regional administrators and chiefs of police up and down the Ethiopian-Somali border refused to continue work and abandoned their posts. They were in fear for their lives, and all moved to Hargeisa and Mogadishu. The men appointed to replace them declined to take up their assignment.

For a long time after this incident, until the signing of a joint peace agreement between the two countries, Somali regions along Ethiopia's border did not have regional administrators or chief of police.

On our side, in the Ogaden, the kidnappings of regional administrators by Somalis and their agents stopped. Free to move about, all the administrators returned and assumed their posts. This problem was, at least for now, resolved. However, we now had another problem demanding action.

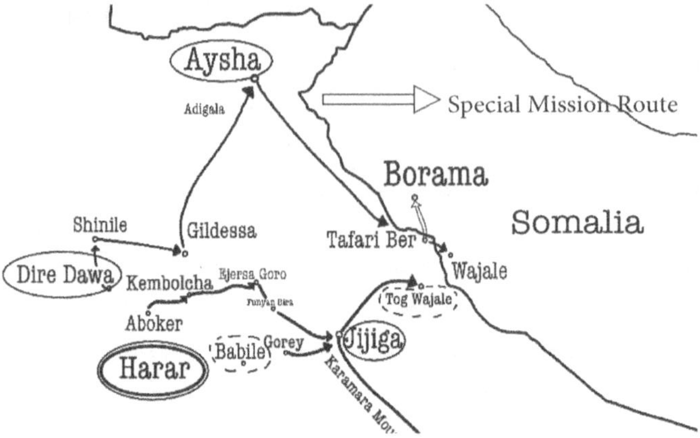

Map 2. The special mission in Borama, 1966

An urgent message arrived for the 3rd Infantry Division Commander from the Ministry of Defense, requesting that a Special Forces soldier who had specialized training in explosives be sent to Addis Ababa at once. The Special Forces Commander summoned me to his office and told me to prepare as quickly as possible because we were leaving for Addis at once. He did not tell me why, and I did not ask. As ordered, I put some articles of clothing in my travel kit and returned to the base. Major Kassa's house was near the compound, so he was also ready quickly. He drove us to Nazareth, where we spent the night. The next morning, we reported to the Head of Operations at the Ministry of Defense, Brigadier General Nega Tegegn. The General knew us and addressed us both by name.

"Your mission in Borama has greatly pleased the Ethiopian Government. We are fully confident that you will succeed in this next mission. Rest assured, the Ethiopian Government will take care of your family should anything happen to you while on this difficult assignment" He then addressed me: "Lieutenant Tesfaye, are you married?"

I answered, "Yes sir."

"Any children?"

"Yes sir, I have one child. He is four-months-old."

"If you are injured, are killed or taken prisoner while on this mission, the government will take responsibility for raising your child, and will provide for your wife. Have no concern about your family." Then, he started outlining our mission to us.

"Throughout the entire border region, neither our military nor our civilian populations, can move without protection. A journey that should take only one day is now a journey of three to four days. There is loss of life on every trip. The attacks and bombings have severely hampered both military activity and civilian life. Our army in Dolo must have its food supplies, and other needs flew in. We are sure that it is only a matter of time before they shoot down the plane. We cannot afford to sit idly by and let this continue. We will take the necessary action to stop these attacks. A decision has been made.

Lieutenant Tesfaye, you have been selected for this mission. You may request whatever explosives you need from the armory. You will transport it by plane to Negele Borena. There, Colonel Gesese Reta will brief you about the particulars of your mission. From Dolo, he will supply you with as many men as you need. Your task is to train these men as quickly as possible. From Dolo, you will take your men into Somalia. Once in Somalia, you will plant anti-vehicle and anti-personnel bombs. You will destroy bridges, police stations, official buildings, paved roads, and troop concentrations. You will inflict as much damage as you can. You will then repeat this in Yod, Elkere, and Imi. I will monitor your progress by radio.

You will await your next mission where you are. In two days, finish your preparations here and get to Negele. Let me know as soon as you complete your preparations. I will supply you with a vehicle to transport the explosives to the airport, and the plane that will take you to Negele."

Then he asked: "Is everything clear?"

I answered: "Yes sir, it is. I accept your order."

"Good. Major Kassa, any questions?"

"Yes sir," Major Kassa said. "I request permission to go with Lieutenant Tesfaye; I don't want him going in alone."

"Fine. In that case, use your vehicle to transport the explosives to Negele."

Major Kassa thanked the General by saluting him.

After receiving the authorizing orders from Brigadier General Nega's office, we headed straight for the explosives depot at Dukem. It took us a whole day to determine the amount and best types of explosives we required for the mission. The next morning, we loaded up the material in our car, gave notice of our preparedness to leave, then went straight to Negele Borena.

Past Yerga-Alem, the road from Addis Ababa to Negele Borena is winding and unpaved, slowing our progress, and when we finally reached the base at Negele Borena, it was getting dark. We let the Brigade commander, Colonel Gesese, know of our arrival the moment we got there. He was expecting us and had made provisions for us.

With armed guards assigned to keep watch over our vehicle and its cargo, we ate supper and then went to bed. The next morning, after breakfast, Colonel Gesese summoned us to his office where he and the Operations and Information Officers all briefed us. After the briefings, we went outside to wait for our plane. While awaiting our aircraft, we walked off the base to stretch our legs and look around. In the near distance, we saw a uniformed guard keeping watch over a good number of camels, cattle, and goats. We approached the guard and asked him to whom these belonged.

"To the army," came the proud reply.

"And where did the army get them?"

"We took them from the enemy," he again proudly said.

"What does the army intend with them?" we asked.

"To divide up the cattle and goats, which will then be slaughtered and eaten. We will sell the camels."

"Does it not displease the people to have their animals slaughtered and sold?" we asked him.

"What people? They have all rebelled against us."

Colonel Kassa and I then both understood that the army was

causing the people to rebel, and discussed this as we walked back to the base.

The plane had landed and was waiting. Following the Brigade Commander, we drove to the airport in our car. We loaded the explosives onto the plane, then trusting the care of our vehicle to the Colonel, we flew to Dolo.

There, the base commander had a vehicle ready for us. We loaded the explosives onto this vehicle and headed to the base. Once we arrived, our cargo was unloaded and stored in a proper facility.

Dolo is a tiny border town. There were perhaps 50 houses in Dolo. Close by, were several small villages. The military base was built near the Genale River which wound like a serpent, silently flowing by the complex. The people who live in the area belong to the Gerimero or Yerer-Barre Somali tribes. They lived in both Ethiopia and Somalia. The Somali government oppressed Gerimeros and therefore, the Gerimeros were often involved in bloody fights with the Somali government. They preferred living in Ethiopia.

Other Somali tribes looked down upon the Gerimeros, and this caused friction between them and the other tribes. We queried the base commander and the Information Officer about their opinions of the viability of using the Gerimero tribe for our mission.

"They can be given the necessary training and will have no problem moving about inside Somalia to set explosives. They have been valuable intelligence sources for us. They live on both sides of the border, and they can use the pretext of visiting of relatives for crossing over the border. If we arm them, they will happily fight alongside our army and make war against the Somali military."

We learned that the Gerimeros were much indebted to the Ethiopian government, and were very loyal to us. We asked the base commander to provide us with twenty healthy young Gerimero men who could speak Amharic and were unmarried or with no family responsibility. These men would have to remain on base until we completed their training. They would not be allowed to leave, not even for a death in the family. They would receive no visitors, and supplies such as sugar, salt, oil, and rice would be flown in from Negele, so they could prepare and eat their meals on base. Those who agreed to these conditions report to us that same day. Twenty men agreeing to these conditions came onto the base and spent the night.

The training started early the next day. As much as possible, we kept the practice skill-centered. For example, how to dig a hole to bury an explosive device. How to hide dirt from where an explosive device was buried by scattering it carefully under bushes so that it would draw no enemy attention. Which part of the explosive device had to face up when it is placed in the hole. How to camouflage the bomb once set in the hole, and other similar skills.

We finished the training in five days. However, the enemy police and government workers along the border received their essential supplies and salaries on certain days of the month. Though we finished our training and preparation, we had to wait another ten days because that is when the vehicles would be on those roads and bridges.

When the fateful day arrived, we selected fourteen of the more capable men. We gave each of them an anti-vehicle mine wrapped in burlap bags. We waited until it after dark, then left Dolo and headed to Luk-Ferede Bridge. This bridge is inside Somalia's border and spans the Genale River. The town of Luk-Ferede is about forty km from Dolo. Moving only at night and staying hidden during the day, it took us two days to get to Luk-Ferede. On the evening of the third day, we buried the mines in strategic locations starting in Luk-Ferede and continuing all the way back to Dolo. We set all fourteen mines. Right before we got to Dolo, one of the mines near Luk-Ferede went off. An hour later, a second one exploded. By the time we got to Dolo, five mines had gone off. These five blasts probably meant that five vehicles were now out of commission.

For a time, we did not know what happened to the other mines. Then on the third day after returning to Dolo, we heard another mine go off. We sent agents to verify the result.

To verify the extent of the injury inflicted on the enemy, we sent out scouts in to investigate. The reports from each returning scout were similar. According to the reports, we blew up five vehicles. The salary for the soldiers and police had been reduced to ashes, along with the truck driver. The blasts also incinerated many of the army's provisions. Counting police and civilians, over 30 people had been killed. Two explosive devices had been found and disarmed by sappers. One of the two mines removed by the sappers went off in Luk-Ferede, in a roomful of officers and lower-rank soldiers were curiously examining it. All the soldiers in the room were killed. The other explosive had been taken in and stored in a munitions depot.

We sent a coded radio message to the Brigade Commander, Colonel Gesese. He sent us a congratulatory note telling us that he already knew of our mission's success through his agents and had already passed on the good news to the Ministry of Defense.

This mission created terror and grief among the enemy. We were preparing to repeat our attacks in Yod and Elkere when the Somali government approached our government through the Foreign Affairs Secretary requesting peace talks. The Ethiopian government accepted the offer. The peace talks revolved around the following key points:

1. Immediate cessation of the use of explosives and mines by both sides.
2. The removal from their secret places, in quick order, of all unexploded devices.
3. Making all closed roads operational.
4. Full protection for sappers and engineers of both sides as they crossed borders and removed explosive they had buried. Prior notification would be given to the military and the police on either side of the border.
5. Complete the removal of the bombs in one month's time.

We had only buried a limited number of mines in Somalia. Most of the mines exploded and others were found and removed by enemy sappers, we had no buried mines to remove. Still, we complained and grumbled that one month was simply not enough time and that we needed more time to remove the many explosive devices that we had buried. The mediating party said that there would be no more time given and that we would have to increase our manpower to finish in the allotted amount of time. We grudgingly accepted the decision. Then, for a whole month, we swam in the Genale River waiting for their side to report its mine removal completion. After receiving it, we too acknowledged that we had finished removing our mines as well.

A few days after completing the mine removal, Colonel Kassa and I boarded a plane and returned to Negele Borena. Our mission lasted three months. From Negele, we drove to Addis Ababa and made our report to the military commanders, and then went back to our base in Dire Dawa.

For some years afterward, the bombs stopped in Sidamo, Bale, and parts of the Ogaden and people were able to move about without fear. Peace had returned to the region. The role that I played during that period to help restore security in that region fills me with spiritual contentment. I believe that I appropriately fulfilled my duty to my country as a professional soldier.

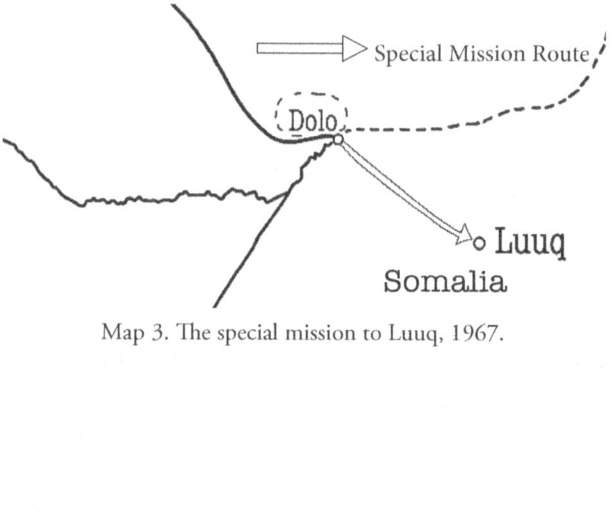

Map 3. The special mission to Luuq, 1967.

CHAPTER 6
Establishment of the Anti-Hijacking Team

In the late 1960s, airline hijackings reached epidemic levels worldwide. Not a day would go by without some news agency reporting another hijacked plane. Tourists, pilgrims, diplomats, businessmen, and other travelers were always in danger of having their flights diverted to hellholes where they would often not have access to water, food, or medical care. Or, to deserts where they could perish from the heat. They became unwitting pawns in political and economic battles. And far too often, passengers with health problems, children, pregnant women, as well as disabled and elderly passengers died due to stress, injury, or lack of proper medical attention. As a result, many travelers, unwilling to risk their lives, canceled their flights and simply stopped traveling by air.

The resulting decline in airline passengers was a severe blow to the entire airline industry. The overall financial impact was undoubtedly significant. But smaller carriers, lacking the resources to ensure their passengers' safety, were especially hard hit and faced the real possibility of going out of business.

The airline that bore the brunt of the hijackings during this period was Israel's national carrier, El Al. They suffered significant losses of property and human lives.

The Israeli government, to protect El Al and reduce the number of hijackings, decided to take aggressive action. Specially trained commandos, disguised as civilians, started flying on all regular flights. Strict protocols were put in place to ensure these commandos remained undiscovered. They traveled like everyone else, using the standard travel documents for their destinations. They boarded and disembarked with the other passengers. Anonymity was their protection.

When it was discovered that El Al had violated the IATA (International Air Transport Association) rule against carrying weapons, there was an outcry by the IATA and the international community. Israel was warned to stop the illegal practice immediately. The IATA stated that if Israel did not comply, their license would be revoked.

The Israeli government stood its ground. Although they were breaking IATA policy, the Israeli government saw no other way to protect El Al, Israel and its citizens. They continued their aggressive anti-hijacking tactics. While Israel's actions were proving very useful in safeguarding El Al from hijackers, Ethiopia's hijacking problems were just beginning.

Various political groups, both within Ethiopia and from outside its borders joined forces to hijack Ethiopian Airlines (EAL) planes. These groups were either opposed to Emperor Haile Selassie's rule or those who sought Eritrean independence such as Shabia and Jebha. Their primary purpose was to cripple the airline economically by preventing it from operating overseas flights. Members of these groups attacked EAL carriers while they were boarding and disembarking passengers, or while the planes were refueling in places like Karachi or Frankfurt. Often, the terrorists would approach them on the tarmac and throw bombs or launch grenades. Sometimes they posed as airport staff in an attempt to get on planes to plant bombs. The attacks on EAL by these groups received international coverage. Passengers hearing the negative publicity and fearing for their lives refused to use the airline.

On more than one occasion, a flight would return from Europe or another continent with only its crew for passengers. This airline, which had been fast gaining in reputation and customers, and which had been a significant source of much needed foreign currency for the nation, was now forced to appeal for help to pay its employees.

The Ethiopian government, meaning the Emperor, wishing to prevent the airline's failure and destruction, and seeing that the problem had to be nipped in the bud before the attackers gained strength and confidence, decided to establish the Anti-Hijack Agency.

The responsibility for establishing an anti-hijacking team fell to the Special Forces commander, Colonel Kassa Gebremariam. Immediately after receiving the authorization, Colonel Kassa summoned a group of us to Addis Ababa where, with little more than our patriotic fervor, we began our task.

We were assigned two to a plane. At the time, EAL had only a handful of planes, but they were kept in constant use with round-the-clock staffing and flew to numerous locations. For example, a flight would leave Addis for Athens. Within forty-five minutes, it would disembark its passengers, refuel, change its crew, and then fly to Madrid. That same day, it would return to Athens, refuel, change its crew again, and head back to Addis. We would remain on the plane the whole time. This process would repeat in Addis, and this time we would take off on a round trip to some West African destination such as Accra. In Accra, the crew would change while we remained for the return flight to Addis Ababa. A single flight security team would switch planes and work for up to forty-eight straight hours without sleep or rest. The flight crews, however, would switch up to four or five times.

There were times when we would have to spend twenty-four hours at a European terminal. We would wait for passengers with connecting flights who were flying in from the U.S., the Far East, and Asia. The airline crew would stay at a hotel in the city when this happened. But, not us. Rain or shine, heat or cold, we would remain close-by to guard the plane. There were times when we were so cold we would lose all sensation in our limbs. We often felt like zombies.

We would go to prison if we were found carrying weapons on the airport grounds. Therefore, whenever we saw airport police approaching, we would move back onto the plane and secure the door. When the airport police moved on, we would get back off the plane and continue patrolling around the aircraft. We would then leave the freezing European continent and head to the scorching heat of Africa. The constant variation of altitude and temperature had a significant impact on our health.

To keep from falling asleep on the job, we frequently went to the toilet and washed our faces with cold water. We would put ice in a cloth and place it on top of our heads. Men who had never touched a cigarette in their entire lives were now chain-smoking. We drank endless cups of coffee. However, when the need for the sleep that had been put off for so long finally broke through, there was not enough ice or coffee in the world to keep nature from taking its course. A human being can stand sleep deprivation for just so long. But how long could we keep this up?. We had come up against a limit over which we had no control. This fact forced us to take another look at how to approach the protection of the airline. We submitted a report requesting an evaluation of the way our mission was being carried out. Colonel Kassa agreed that the matter needed additional study, and arranged a meeting with the Director of EAL and the Civil Aviation Director. The meeting focused on the following points:

- The need to increase the number of men on the anti-hijacking team.
- The establishment of an anti-hijacking training program to increase the availability of qualified personnel.
- A program to recruit qualified men from all branches of the armed forces and provide them the specialized training before their deployment;
- Allow the Flight Security Agents to get adequate rest between shifts by having them change and rest at various stations in the same way the regular civilian flight staff did.
- Secure additional protection for our planes from the countries and their airport authorities into which our aircraft would fly.

- The need to recruit ground agents and station them at the various airports utilized by EAL. These agents would pose as regular civilian employees of EAL but would be trained to screen suspects and disarm explosive devices should they be discovered.
- Establish closer working relationships between Flight Security Agents and airline employees.

The ideas presented were discussed at considerable length and resulted in some positive outcomes. At the end of our discussion, I was tasked with setting up the training program and become its director.

We had to augment our current staffing until the graduates of the new program started taking their posts. It was decided to supplement the Flight Safety Agents by temporarily taking additional men from Special Forces and Airborne units.

I lost no time in setting up the training program and selected a number of good trainers. After we established the recruitment and training guidelines, the branches of the Armed Forces were instructed to send us the required number of suitable candidates.

We administered an entry exam to test whether or not the candidates sent to us by the three branches met the established criteria. Accepting only those who passed the test, we immediately began the accelerated two-month training program. Our goal was to provide this first round of trainees with the skills needed to neutralize a hijacker with one shot, to defuse explosive devices, as well as the hand-to-hand combat techniques to stop aggressors in close quarters.

Not long after our first class had graduated from our training program, an EAL Boeing 707 en route from Madrid, Spain to Addis Ababa was hi-jacked before the en route stop in Athens. A team of our new graduates was on the flight. They remained calm, followed the hijacker's instructions, and waited for an opportunity to strike.

The hijackers knew that trained flight security agents were aboard the craft. However, they didn't know how many or who they were. The hijackers thoroughly searched all the passengers, the aircraft, and luggage. They tried threatening over the public address system, stating that they knew that armed men were aboard and that they should hand over their weapons immediately. The hijackers had no luck turning up any weapons during their search because the security agents had hidden them very well.

One of the hijackers placed his gun against the head of a female flight attendant and forced her towards the cockpit. He was going to use her to force the pilot to unlock the cockpit door and then have the pilot change the flight destination to their chosen location. The hijacker pushed the woman before him holding a gun against her head. One of the security agents stationed in the first class section got up very quietly from his seat as soon as they went past him. As they continued toward the cockpit door, he fired his gun, striking the hijacker in the back of the neck. Passengers saw the man fall to the floor but heard nothing. The agent's weapon was fitted with a silencer which muffled the report of the pistol. The second agent, seated close to the front of the economy section, took his cue from the team leader and moved into the first class section where he shot the hijacker addressing the passengers with the PA system through the forehead. The other two security agents towards the back of the plane shot the third hijacker. Not one of the hijackers managed to get a single shot off. The explosive device rendered safe and removed, and the aircraft continued on its scheduled path. Not one passenger was injured, and there was no damage to the plane. The Flight Security Agents handled the entire incident quickly, smoothly, and professionally.

The passengers were impressed by the speed and the efficiency of the security officers. When the incident became publicized, there was widespread support from many countries. However, a few countries, and the IATA were critical. The condemnation that had previously been leveled at EL AL, streamed in against Ethiopia, claiming that this kind of act should not have taken place at the risk of passengers' lives while it was possible to negotiate with the hijackers. Like the Israelis, we ignored the condemnation and criticism and quietly maintained our policy.

The hijackers, not learning their lesson from Athens, tried a second time. This time they hijacked a DC6 plane en route from Addis to Djibouti with a stopover in Dire Dawa. Landing in Dire Dawa, some passengers disembarked, and other passengers going to Djibouti boarded. While the plane was in flight to Djibouti, the hijackers announced to the passengers over the PA system that the flight had been hijacked and was being diverted to Aden in Yemen. On that particular flight, due to a shortage of personnel, there was only one flight security agent on the plane. Since there were two hijackers, the agent decided it would be safer to take action when the aircraft was on the ground. Therefore, he waited until the plane landed at Aden airport.

As soon as the door opened and the ladder was put down for passengers to disembark, the security agent slipped out and took up a firing position. He waited for the two hijackers to exit the plane and shot them dead. The airport police arrested our security agent, charged him with disturbing the peace, and threw him in jail. The aircraft, however, returned to Addis Ababa, safe and sound. After two months of negotiations between the Ethiopian and Yemeni governments, our flight security agent was released and returned home.

During this time, we continued training recruits drawn from the three forces and slowly began to fulfill our operational needs. Each flight was now assigned four flight security agents. To the task of ensuring the safety of the plane and its passengers, we typically placed one security agent in the first class section, in the seat on the aisle. The second security man usually sat in the second class section, in the first seat of the front row on the right side where he had a good view of the cockpit. The third security agent typically sat in the middle of the second class section across from the second security man, and the fourth security agent sat on the right side among the seats at the back of the plane. They formed a zigzag so they could see each other and were given aisle seats to enable them to move about swiftly. Security agents were also stationed in airports to help protect passengers. These agents would often tip the flight security agents when they detected something suspicious. They could also defuse bombs if necessary and would inspect luggage for possible hidden weapons. We established a good working relationship with the personnel of the Ethiopian Airlines.

While our teams gained a lot of experience from performing their duties, the hijackers did not seem to be able to learn from their mistakes, and soon made a third hijacking attempt.

The third hijacking drama began when a flight from Addis to Europe took off from Bole airport and began its ascent to its flight altitude. The hijackers suddenly sprang up from their seats and brandishing hand grenades and pistols, took over the plane. One of the hijackers seized the PA system and repeatedly announced that the flight was under the hijackers' control and that everybody must carry out any orders given by the hijackers or face serious consequences.

Placing their index finger in the rings of their grenade, they threatened to pull out the safety pin and detonate the hand grenades, wiping out the passengers and the plane as well as themselves. They then ordered all passengers to put their hands on their heads and remain seated and silent.

One of the hijackers started struggling with the door of the cockpit in an attempt to get in. The flight security agents decided to make their move.

The flight security agent in first-class shot the hijacker who was struggling with the door of the cockpit. The second agent began to scuffle with the hijacker using the PA system attempting to wrestle the hand grenade from him. He quickly got the better of the hijacker. The two agents at the back promptly took out the rest of the hijackers. It did not take long. The flight security agents soon had control of the aircraft. Once again, they returned to land in Addis with no injured passengers or damage to the plane. This time, when the story that we had foiled a third attempted hijacking went public, we received a great deal of praise. Even the IATA was beginning to realize that these hijackers would continue to be a plague on international air travel and that it could not be stopped through negotiation. We had become international heroes.

Putting an armed flight security agent on a commercial aircraft used to be considered criminal. However, after our successes, several other countries who were dealing with hijacking problems invited us to share our experience with them. They too wanted to safeguard their passengers and aircraft with armed security agents.

Many who were once afraid to fly on Ethiopian Airlines now would not use any other airline because the Ethiopian airline had earned such a reputation for having such effective security. The same Ethiopian Airlines that had recently faced bankruptcy, thanks to the excellent work of our flight security agents, regained its good name and was now getting more passengers than it could handle, and was soon safe from bankruptcy and had become profitable again.

After serving my country and the Ethiopian Airlines from 1969 to1972 as the director of the anti-hijacking team, I requested a transfer back to my unit in Dire Dawa. Then, in 1973, I was sent to the United States of America for Special Forces training. After completing this training, and even though I was no longer officially part of the anti-hijacking team, I continued to help train the recruits sent to Dire Dawa until 1975.

With allied students in the Special Forces training
Author: 3rd row, 3rd from Left.

With members of the Special Forces ready for a jump
at Fort Bragg, North Carolina, USA. 1973
Author: Front row, 1st Left

Pictured with our operation officer, 1973
Author: Back row, 1st Left

Pictured with Special Forces commander, 1973
Author: Back row, 1st Left

Pictured is the family that sponsored me during my training in North Carolina: Captain and Mrs. Lents, 1973

I received a letter of appreciation from the commander of Flight Security, Lt. Colonel Kassa Gebremariam, for my contribution to the training of the flight security men. The letter reads as follows:

October 22/1965 (E.C)
No.1//100/9/488/65

To: -2609 Captain Tesfaye Habtemariam

Letter of Appreciation

You won the admiration of the officers and NCO's drawn from the ground force, air force, and naval force whom you recently gave special military training to ensure the flight security of the Ethiopian Airlines for your determination, wisdom, capability and volition. I, too, followed the training closely and admired your efforts and your moral strength. I feel proud, therefore, to inform you that I had this letter of thanks be attached to your personal file in appreciation of your volition and your love of your unit which I hope will continue.

Lt. Colonel
Commander of the special military training center
/1260 Kassa GebreMariam

CHAPTER 7
The 1974 Popular Revolution and the Armed Forces

During the Haile Selassie era, each branch of the military, as well as the specialized units, had their own unique uniforms, caps, and emblems. Every unit was proud of their distinctive unit uniforms. During peacetime, military leaders used them to maintain discipline and esprit de corps. During times of war, these symbols were a source of pride, representing ideals such as courage, determination, and heroism. Members of the Ethiopian military attached great significance to these symbols and in combat, would fight to the death to keep these physical representations of their deepest commitments out of the hands of the enemy.

After the rise of the socialism in the country, the new government, the Derg, announced that all the armed forces would wear the same caps, emblems, and uniforms. The uniforms of officers and non-commissioned officers would all be the same. This announcement created great dissatisfaction among all members of the armed forces because it deprived them of their pride in being able to represent their distinct identities and abilities. Many military units had little choice, however, but to swallow their dissatisfaction and follow orders.

The airborne troops, however, were a different matter. They felt very strongly about this after so many sacrificed their lives on the battlefield to prevent their prized red berets from falling into the hands of the enemy. Ordering the airborne soldiers to abandon their red beret was like telling them to divorce their beloved wives.

The airborne troops requested that the Derg reconsider the directive concerning their berets. The Derg responded that the airborne soldiers should first accept the order and turn in their red berets, as ordered. Once they followed the directive, they could argue their case. The airborne troops continued to defy the order. They began to shout the slogan, "our red berets or a bloody fight" to anyone who could hear it.

When the Derg realized that the airborne troops were serious about the issue, it decided not to escalate the dispute. Some members of the Derg began letting the airborne soldiers believe that their request to keep their red beret would soon be approved. The battalion commander and some troops, believing what these members of the Derg told them, wore their red berets even when visiting the Derg headquarters.

When the Derg members pretended not to notice, the airborne troops were further encouraged by this seeming lack of scrutiny. But the Derg was deeply concerned that if the airborne forces, with their insubordinate attitude, remained at their base in Debre Zeyit, they would set a bad example for other military units. Therefore, the Derg gave the airborne troops a mission in Bale and used this as an excuse to move the airborne headquarter from Debre Zeyit to Bale.

The new Airborne unit recruits for 1975 were ordered to accompany the unit to Bale even though they had not yet completed their parachute jump training. The recruits questioned whether it would not be more appropriate to complete their training before going on the campaign. The fact that the Derg summarily dismissed the inquiry and the recruits were told to comply with the order, intensified the dissatisfaction of the airborne units with the Derg. It also opened a way for opposition groups to begin to form. The opposition groups began to fan the flames of discontent both from within the military and without.

They started with the recruits.

Now, if you die, how are people going to lament your death? Are they going to grieve the death of an airborne soldier? Or else, are they going to mourn the death of an infantry soldier? You will not be seen as an airborne soldier because you have not completed your training. You will not be allowed to wear the red beret or the wings. That means that you are not an airborne soldier. After only four jumps, with one remaining to receive the red beret and the chest wing emblem, you were ordered on a mission. You have chosen silence.

They then moved on to the airborne soldiers.

You brought the Derg to its knees, haven't you? You are indeed heroes. Courage! Don't abandon your red beret. While others chose to be cowards and take off their berets, hiding them in a box, you stood up for your rights and continue to wear your red beret.

It was because of this agitation for rebellion that the Derg sent the airborne units on the operation in Bale. After the airborne unit left for the Bale region on their mission, the agitation for rebellion continued.

In the meantime, the Eritrea situation was getting worse. Therefore, in of June 1976, the airborne unit was ordered to move from Bale to the Eritrea region.

As soon as we received our orders, we traveled by vehicle, covering the distance between Bale and Debre Zeyit, a trip usually taking three days, in only two. As soon as we arrived at Debre Zeyit, I asked permission to go to the countryside to attend to an urgent family matter. The airborne troops remained in Debre Zeyit until air transport to Asmara was ready. During the short stay at Debre Zeyit, the conspiring against the Derg continued.

When the airborne units arrived in Bole International Airport on their way to Eritrea, they had planned a coup. An advance company was ordered to go straight to Meshwalekia, enter the compound of the 4th Infantry Division, and put the Derg members under its control. A second company had orders to take control of the radio and television stations, telecommunication companies, and banks. They also had orders to block the main roads leading into or out of Addis Ababa, as well as to place under their control troops still in their camps. As soon as these actions were complete, an announcement was to be made to the Ethiopian people over the radio and television stations, that the Derg had been overthrown.

As the advance company continued its way to Addis and reached Dukem (35 kilometers from Addis), the company commanders received their final orders for the coup attempt. However, the Commander of the advance company, for reasons that are not clear, instead of heading to Meshwalekia upon arrival at Gotera, headed to Bole airport via Wello Sefer together with his troops. From the airport, he and his troops entered the plane waiting for them and proceeded to Asmara. The second company, realizing that the plan had failed, had no choice but to head to Bole airport and, following the advance company, fly to Asmara.

While their initial plan to overthrow the government failed, the airborne troops arrived at Asmara and continued their plans using secret communication between themselves, Brigadier General Getachew Nadew (a former airborne unit commander), the Air Force, and other military branches.

The commander of the 2nd Infantry Division, having learned of the conspiracy to overthrow the Derg by the airborne units, decided to counter this by moving the airborne units away from Asmara.

The Asmara to Dekemehari road had been closed to motor traffic for an extended period. The closure was because of the severe risk to human life and property resulting from the guerrilla warfare and anti-vehicle mines buried by the separatists in Eritrea. Supplies were currently being airlifted into the soldiers by helicopter. The airborne forces, together with the 31st Battalion, were tasked with freeing the Asmara-Dekemehari road from the enemy and clearing the mines. I ended my leave and arrived in Asmara to resume my command before this mission began.

I knew nothing of the conspiracy. We went into battle the day after I arrived from Addis, and the commander did not take me into his confidence. It may be that he had not had time to talk to me. Or, he may have decided not to inform me because I was new to the unit. After receiving our orders, we advanced from Asmara toward Dekemehari. After moving forward about ten kilometers, the fighting began.

The separatists had been told by their leaders that the airborne soldiers from Debre Zeyit were raw recruits; that they had not had any airborne training and had not seen combat. Their commanders told them that the red berets and the chest wings we wore were fake. So emboldened, they sprang up from their fighting positions and charged us as soon as the shooting began. They sought to overwhelm us and make us prisoners of war. However, we were not simple recruits. We waited calm, poised and ready. When they came into range, we shot them down.

The Asmara-Dekemehari road, closed for months, was now cleared of mines and guerrilla fighters and was again open to traffic. Trucks again began to deliver supplies to Senafe and then return to Dekemehari without any incident.

Having accomplished our mission, we prepared to return to Asmara from Dekemehari. We received a telegram message from the Commander of the 2nd Infantry Division, ordering us to stay where we were until we received further instructions. When our commanders asked for an explanation, they were told that the separatists, though badly hurt in the battle we had just fought, they were already regrouping to attack again and therefore, it would be safer to remain at Dekemehari for the time being.

Our commanders responded to the Commander of the 2nd Infantry Division through our battalion commander stating that this was ridiculous. We were better armed and trained than the separatists and had nothing to fear. It was imperative that we were allowed to return to Asmara.

Realizing the delicate nature of this situation, the Commander of the 2nd Infantry Division permitted us to return to Asmara. He had come up with an alternate plan. The next day we left Dekemehari. We were halfway to Asmara when there was an announcement on Radio Ethiopia that Brigadier General Getachew Nadew, the overall Commander of the Army in the Eritrean region, had been executed. We were shocked to hear about the death of the Brigadier General. He had been the commander of the airborne unit for ten years and was a beloved leader, deeply respected by his men.

Brigadier General Getachew Nadew, as commander of the airborne units, had been granted considerable leeway in drafting any officer or NCO from any military unit that he considered to be capable for the airborne. He could dismiss anyone he did not feel met his standards. He could also appoint anyone to fill a vacant office and select anyone for training at home or abroad.

When he was appointed as the administrator and the overall Commander of the Army of the Eritrea region and left for Eritrea, he took his personal staff from the airborne unit with him.

The respect the airborne units had for Brigadier General Getachew Nadew was deep. The failure of the planned conspiracy and the resulting death of General Nadew created a shockwave. Many felt that his supporters would have to account for the actions of the conspirators. The execution of General Nadew was an unexpected blow to his followers. Their plan to assume power shattered and now they were dangerously exposed. They had to find a way out.

After our return to Asmara, the conspirators began to feverishly work at including the Air Force and other military branches in the plot. The Commander of the 2nd Infantry Division, who was aware of the scheming, ordered the airborne units to move to Mendefera and relieve the 30th Infantry Battalion.

The primary task of Airborne troops is to serve as a reserve force. They should not be deployed in trench warfare. We knew that the command given to us did not make tactical sense. However, since we were in a war zone, we had to accept and carry out our orders. Refusal to obey orders was a serious offense, punishable by death during times of war.

When we moved out of Asmara, the battalion commander ordered me, upon arrival at Mendefera, to occupy the H.Q of the 30th Battalion together with my company. Upon our arrival, the battalion commander, Lieutenant. Colonel Alemayehu Asfaw, approached me accompanied by several of his men. All stood with their weapons pointing at me.

The Lieutenant Colonel Asfaw asked me angrily, "Why did you enter the headquarters with your troops? Who told you to enter here?"

I replied to him, "You did."

"We shall talk about it later," he said. "Now, move your company to the troop assembly area just outside the southern part of the town," he commanded.

Obeying the order, I moved my company to the location specified. I had no idea what was going on, and no one explained the situation to me.

The 2nd Company stood in a line along the right side of the road to Addis. This road divides the town into two halves, and after crossing the Mereb river, proceeds through Adwa and on to Addis. My 4th company was about to line up behind them when they told us to continue forward. When we moved ahead of the 2nd Company and stopped, we were told to keep going.

The area ahead of us was enemy territory; I did not understand precisely where we were supposed to keep going. I went to the battalion commander and asked him, "Where are we going?"

The battalion commander, Lt. Col. Alemayehu Asfaw, responded, "Haven't you heard the news?"

"I heard nothing; what is going on?" I asked.

"Our men that remained in Debre Zeyit were driven out of their camp by the Derg using tanks. They were chased out to Tasse mountain. Radio communication has been disrupted. So, we are going to Debre Zeyit because our men here want to know what has happened to their families," he said.

"How can that be?" I asked. "Do we have orders to proceed to Debre Zeyit?"

"Orders from who? What government is there? You call the Derg a government?"

"Yes, of course.," I replied.

"Forget the Derg."

"I beg your pardon, Colonel. Alemayehu. I don't think this is right. If we travel one kilometer away from Mendefera without orders from the government, the government will consider us as enemies from that point forward. We will not be any different from separatist in their eyes. At any rate, how are we going to fight against the fighter jets, tanks, and heavy guns of the government for the 1000 kilometers from here to Debre Zeyit?"

"No problem whatsoever" he replied.

"If that is so, I am not going," I told him, "I have orders to safeguard Mendefera, and I shall do so."

"Our whole unit is going. How are you going to stay here all by yourself?"

"I don't think my company will abandon me," I said to him and proceeded to my troops to inform them of the situation.

"The choice is yours," I said to them. "If you choose to fight against the government from here to Debre Zeyit, go get into the trucks. If, on the other hand, you believe we should follow orders and complete the mission given to us by the government, then line up behind me."

While a few of the recruits headed towards the vehicles, the seasoned soldiers all rallied behind me. Many of the older soldiers from the 2nd company came to join us as well.

When the battalion commander, Colonel Alemayehu Asfaw, and the other initiators of the conspiracy saw that most of the troops rallied behind me, they changed their minds.

"Let's move back, and camp on the Catholic Church grounds," suggested the Colonel. We went to the Church grounds and set up our camp.

Our unit was now divided into two factions. Members of Colonel Alemayehu's faction considered me an enemy because I had foiled their plans. I soon got warnings that they were planning to kill me. The old-timers who rallied behind me, identifying the conspirators, prevented them from approaching our area in the camp. Both factions carefully observed each other the entire night.

The next morning, we got hit with an unseasonable rain. About forty young officers and NCOs, all leaders of the conspiracy, took advantage of the cover provided by the rain, and with their rifles and equipment, slipped down the valley to join the separatist group. It appeared to me that they had prior communications with the separatists.

I reported the desertion to the 2nd Division and re-organized the remaining troops. Establishing command and control, we continued our task of overseeing Mendefera.

An investigation committee led by Major Berhanu Bayih flew in by helicopter from Addis Ababa. Those conspirators who had not yet defected and had remained behind think that they might escape detection were identified and taken to Addis for further questioning.

These soldiers were jailed without a trial for eight years. It was later rumored that Colonel Alemayehu had somehow made it to the United States, while the fate of many of the other conspirators remained unknown to us.

While the airborne unit was assigned to oversee Mendefera, the troops of the 2nd Division were deployed in small units all over Eritrea. The distances between each unit made it hard for one to give aid to another if needed, or to complete their task of safeguarding a particular area.

The widespread deployment of the soldiers of the 2nd Division made it easy for the separatists to concentrate its forces and attack and eliminate each detachment, one by one.

One of the isolated detachments placed in the western desert was the 15th Battalion, stationed at Nakfa. The separatists repeatedly attacked the 15th Battalion trying to wipe it out and strengthen their position. However, the 15th Battalion put up a fierce fight each time.

The 15th Battalion's strength was steadily being reduced, the dead and wounded resulting from the daily attacks by separatist elements. They were also facing a critical shortage of food and ammunition. Despite this, they stood their ground, counter-attacking often. Finally, getting desperate, the 15th Battalion put out a call for aid to the entire Ethiopian military.

Here is the text of the telegram message sent by Major Mammo Temtime, the Commander of the 15th Battalion, on October 17th, 1976, to the entire Ethiopian military:

> "Two hundred of us have fought as hard as we can, day and night, for one whole month against separatists with much greater numbers and weapons. We intend to keep fighting until our last breath. Over 60 of our wounded men are suffering, rotting in the trenches. Our concern is not for our lives. We are concerned about having the weapons and equipment in our possession taken by the enemy. And, defiling the good name and reputation of our country. Since we have not received the aid we requested from our comrades while fighting all this time, we are beginning to doubt that we have a government that we can rely on and in which we can have faith. If we do have such a government and comrades who are concerned about our country and their fellow soldiers, we should be given immediate help. Not with words of support, but by action. This telegram is our final message."

When this telegram was passed by radio communication to all members of the armed forces, the Airborne troops also received the message. The officers gathered together to discuss this request for aid. We felt that this concerned us, primarily since this was the kind of mission for which we existed. We requested by telegram that the Commander of the 2nd Division permit us to parachute in and reinforce the 15th Battalion. In a short while, we received a response from the 2nd Division telling us to be ready. That night, we selected a total of ninety-nine soldiers who had completed jump training and organized them into platoons and companies. The next day, we flew by helicopter from Mendefera to Asmara to complete jump preparations.

CHAPTER 8
Discussion about the Nakfa Campaign

When we arrived at Asmara airport, Major Shibabaw Zeleke together with the HQ officers of the 2nd Division, came to the airport to welcome us. From the airport, he followed us to our assigned quarters. I had no prior acquaintance with Major Shibabaw, and this was the first time we had met in person. I knew that he had been the commander of the 30th Battalion. He approached and told me that he had been assigned to command the airborne battalion and that we should work together from now on, consulting each other as needed. I said that this was fine with me.

Major Shibabaw had started as a private in the Airborne. Then he joined the Holeta Cadet Training School and became an officer. When he returned to the Airborne, he had attended a training course on intelligence and worked as intelligence officer of the Airborne. Those who knew of his work as an intelligence officer said that he was a very suspicious man who did not trust his own shadow. This meant that when he said we would work together, consulting each other, he was trying to determine where I stood. When we were doing ground assault training, he followed our exercises closely.

While we were carrying out pre-jump drills, the 2nd Division commander called for a briefing. Those invited to the briefing were staff officers of the 2nd Division, the commander of the Asmara Air Force base, the operations officer of the base, the political and security chiefs, the commander of the Airborne Battalion, and me, now the commander of the Expeditionary Force.

The commander of the 2nd Division began his briefing on the general situation, the relative strength of our forces versus the enemy, and the activities that had transpired.

Some units, unable to stand up to the enemy, abandoned their defensive positions hurriedly, leaving their wounded, weapons, and equipment to the enemy. They either rejoined other government forces nearby or surrendered. The 15th Battalion, on the other hand, despite the continuous assaults by separatists over the last month, was still holding their position and fighting heroically. Although the government had made several attempts to break through the enemy lines and relieve the 15th Battalion, so far, they had not been successful.

The government was now preparing a major drive utilizing a sizable ground force reinforced by tanks, heavy artillery, and close air support. The preparations would take some time. Until then, the government had decided to have the airborne unit parachute into Nakfa to assist the 15th Battalion and keep it from being wiped out.

The problem we had was to decide how best to drop the Airborne forces in to help the 15th Battalion without suffering too many losses while doing so. Our fighter jets needed to provide air support to the 15th Battalion in order for the Airborne forces to land safely. However, we all knew that the area around their position at Nakfa was protected by different kinds of anti-aircraft weapons with various ranges, including the Soviet-made Zu-23.

The Zu-23, a 23 mm twin-barreled auto-canon, has an effective range of 2,500 meters. A transport plane would need to fly at an additional 500 meter altitude to be entirely out of the Zu-23's effective range. Any distance lower than that increases the chances of being hit by the anti-aircraft weapon, putting both plane and the airborne soldiers in great danger.

In contrast to this, the higher the altitude the drop was made from, the longer the men would be in the air and vulnerable to being picked off by the enemy as they dropped. Ideally, from the time the parachute opens to the time they hit the ground, they should not be in the air for more than 1 to 2 minutes. The briefing discussion tried to reconcile the dilemma, but by lunch, no solution was found.

The commander of the 2nd Division consulted his watch, and announced, "It is now lunchtime. We shall break here. The briefing will resume at 1400 hours." He stood up and left the briefing room. We all headed back to our camps for lunch.

When we arrived in our area, and I got out of the vehicle, my men eagerly gathered around looking for news on the jump. I explained the difficulty of the situation. The men were adamant and unified.

"We are ready to jump at an altitude of 3,000 meters, and if necessary at an altitude of 5,000 meters. Pass this message, on our behalf, to the commander of the 2nd Division."

"Ok, settle down. Let me eat my lunch. Discuss the matter among yourselves, and we can talk again after lunch." As I ate lunch, different ideas were swirling around in my mind.

From the intelligence officer's report and the map, it was evident that the landing zone was very narrow and encircled by anti-aircraft batteries, mortars, machine guns, and small arms. I didn't see how it would be possible to jump 2,500 meters, reach the ground alive, and join the 15th Battalion. Part of me was adamant. It is not proper to accept an impossible mission as though it were possible. Unless I was forced to do so, I didn't want to take this mission.

On the other hand, I thought about my duty as a soldier and a leader. If ordered to, I would accept this mission and carry it out to the best of my ability. It was, after all, my job. Wrestling with these contradictory ideas, I finished lunch. I had no idea what I ate.

When I returned to talk to my men, they were already waiting for me. The company sergeant major, after calling the soldiers to attention and saluting reported that they were all here, ready for the discussion. I returned the salute.

"Let them sit in the shade," I ordered. The company sergeant major led them to a shady place. Approaching the men, I asked, "What have you decided?"

"No change; we stand by what we said earlier. We don't want the mission to be canceled. Let us do our duty. Because of the men who defected from our midst, we are all considered outcasts by the Ethiopian armed forces and by the public. We do not feel we deserve this. We prefer to make history and die rather than live with the shame.".

"Those of you who are willing to jump by parachute at an altitude of 2,500-3,000 meters, raise your hands," I said to them. They all did. "All right, I shall tell the commander of the 2nd Division how you feel." I headed back to resume the briefing.

While we were heading to the division's briefing room, I could not reveal the decision of the troops to Major Shibabaw. His driver and his two bodyguards were in the car with us. As soon as we reached 2nd Division headquarters, and got out of the car I told him, "The troops want to jump."

"That is crazy, "he replied. "Not possible by any means." Before we could discuss the matter any further, we were ushered into the briefing room.

In the briefing room, we took our seats, senior officers front, the rest of us at the back. As I was the most junior of the officers, I sat on the last chair, behind Major Shibabaw.

Upon the arrival of the commander, we all rose from our seats and stood at attention. The deputy commander reported to the commander that all the participants in the conference were present. The division commander acknowledged the report, then told us to sit.

According to senior government officials, rescuing the 15th Battalion is imperative. And the 2nd Division Commander reminded us that he was not an expert in airborne operations. He told us that this mission would not be canceled under any circumstances. Therefore, it was up to officers of the Air Force and the Airborne Battalion to come up with a solution to this problem.

The discussion around the table was the same as it was before lunch, with the same objections and problems. No solution seemed close to being found.

The silence was heavy. I sat there arguing with myself, torn between expressing the feelings of my soldiers and not. Finally, I decided that I had to speak up, and raised my hand for permission.

The Commander of the 2nd Division knew me by name. "All right, Captain Tesfaye, Commander of the Expeditionary Force, go ahead." I sprang up from my seat to attention and gave a military salute, thanking him for giving me a chance to talk.

"I have listened carefully to the expert views of both the Air Force and the Airborne commanders. Being a paratrooper, I am perfectly aware of the dangers faced by my soldiers jumping from the altitude required for this mission.

On the other hand, our conscience will not let us turn a deaf ear to our heroic brothers of the 15th Battalion who have been in a two-month long life-and-death struggle and have now asked for our help. If we delay in assisting them any further, whatever reasons we may have to do so, neither our conscience nor history will spare us harsh judgment. This mission concerns the Airborne Battalion explicitly. Moments like this are why we train. This mission is our job, and if we do not do it, we risk losing our self-respect. If we fail to respond to this call for help, we know that we will be found guilty of negligence and cowardice by history and future generations. We have no desire to see ourselves in such a light. It is better to make any sacrifice today than risk the danger of offending our consciences, justice, and history.

Knowing that the 15th Battalion's request for aid is ours to heed and fully aware of the danger we face in jumping from such an altitude, it is for love of our brothers-in-arms and our country, that we have decided to parachute into Nakfa and fight alongside our brothers of the 15th Battalion. If we fail in helping them break the siege, we shall at least die alongside them. We are ready for the danger. Even if we are all killed while still in the air, I want now to make it clear to the Commander of the 15th Battalion, the Commander of the 2nd Division, to the Armed Forces and the Ethiopian people, that we have decided and are ready to jump. I thank you sincerely for giving me the opportunity to speak." I saluted and resumed my seat.

The Commander of the 2nd Division smiled. A significant burden was suddenly lifted from his shoulders. "Captain Tesfaye, we thank you for your words as the representative of the Expeditionary force. I don't think there is any need for any further discussion after hearing what you said. The discussion ends here. We will now proceed to develop a plan of execution."

The mood of everyone in the hall changed when I finished talking. Most especially the Commander of the Airborne Battalion. He was furious that such a decision was made and expressed without his prior knowledge and approval. Major Shibabaw raised his hand, and being recognized, spoke.

"As far as I know, I am the commander of the Airborne Battalion. If some soldiers have made such a decision, they should have brought it to me to convey, not to Captain Tesfaye. I want it understood that the words just presented by Captain Tesfaye, are those of a few renegade soldiers and Captain Tesfaye himself. I will not be responsible for the disaster that will befall these men."

"Major Shibabaw," the 2nd Division Commander replied. "there is one thing that you have not grasped. The message passed by Captain Tesfaye is that of the Expeditionary Force under his command. You should realize that he is neither interfering in your leadership nor challenging your authority." The Commander paused to make his point, before going on. "We will not turn the briefing room into a forum for arguments. The time we have left is very short. We must proceed to the next step." The briefing was over.

We were ordered to remain at the Officer's Club while the Operations Officer prepared orders for both the Air Force and my Expeditionary Force. All officers involved were required to be present. About an hour late, the Operations Officer completed our orders, and we returned to the briefing room to receive them.

The operation order of the 2nd division

After ascertaining that we were all present, the operation officer began to present his operation orders to us. The Operations Officer covered the general situation and events leading up to the current situation as well as what would likely happen if the 15th Battalion fell and Nakfa fell into the hands of the separatists.

He also covered the known strength and disposition of the enemy forces we would encounter during the operation as of the last intelligence reports, including any anti-aircraft weapons and other heavy weapons the separatists had.

Next, the dire situation the 15th Battalion was in was covered, as well as the planned relief column currently making preparations to get underway. We had heard much of this before. It had all been covered in our earlier briefing. I was waiting to listen to our orders. Finally, the Operations Officer got to them.

While this brigade completes its preparation, it has been decided to send in an airborne unit, to parachute into Nakfa and help the 15th Battalion hang on.

Topography: There is a mountain range running from north to south along the western side of the military camp at Nakfa. In the areas separated by the mountain range, there are hills of modest height covered with sparse thorny bushes and short grass. These hills can be used as cover.

Mission: The Airborne Expeditionary Force will jump from two C-119 airplanes provided by the Air Force over the military camp at Nakfa and reinforce the 15th Battalion. At the same time, one Pathfinder will insert from a helicopter.

At the time of drop, the Air Force fighter jets will provide air cover to protect the parachuters, the transport planes, the Pathfinder, and the helicopter.

D day will be October 31 and H hour will be 0700. The logistics officer of the 2nd Division will fulfill equipment and logistical requests made by the Expeditionary force. All the preparations are to be completed by October 30 at 1500 hours. Completion status will be reported directly to me.

Command and Communication

Communications between the Air Force and 2nd Division headquarters for the duration of this operation will be by telephone and radio communication. Before the jump, communications between the Airborne units and division headquarters as well as the air force will be by telephone or in person. After the jump, the Airborne units will establish direct communication with the 15th Battalion and make use of encrypted short-range radio communications.

2nd Division will supply transport for the Airborne company, together with its parachutes, to the airstrip before their scheduled take-off time.

After receiving our orders, we all went about making our preparation. The Airborne Battalion Commander and I returned to our military camp without saying a word to each other. He was still furious with me.

When we got out of the car, he called, "Captain Tesfaye!"

"Yes, comrade" I replied, approaching him.

"You will prepare the operation order since you started this in the first place. And. you will also give tomorrow's briefing." With that, he left.

Major Shibabaw was the commander, not me. My job is to carry out the orders given to me by him. I was a bit flabbergasted when he said I was to present the briefing. Putting that aside, I called for the two platoon leaders and the sergeant major.

"Have the men make the necessary preparations" I ordered. They would see that the men drew the required weapons and equipment. Also, I wanted three basic loads of ammunition and four hand grenades for each soldier as well as three basic loads for the crew-served weapons. As of now, the men were restricted to base.

I next made my way to the Division Operations Office to prepare our operation orders. In the office, the battle map on the wall provided all the information I needed to complete my orders.

I began writing the detailed operational plan for the Airborne

Battalion Expeditionary Force and its assigned task of reinforcing the 15th Battalion at Nakfa.

I went back to our assigned area and finding the two platoon leaders and the sergeant major there; I asked them what preparations were complete. They reported that all the requested weapons and equipment was drawn and distributed to the men and that they had inspected each soldier in person. Everything was ready.

On October 29th at 0700 hours, I ordered the men to fall in for inspection with their full combat loads. Acompanied by the two platoon leaders and the sergeant major, I began the review. Each soldier was fully equipped and his equipment correctly secured so it would not come loose or create unnecessary noise. Aside from a few minor adjustments, they were ready.

I thanked my soldiers, their platoon leaders, and the sergeant major for their great work in completing the preparations. I then ordered the men to relax, but to stay close to the camp. I then headed off to find Major Shibabaw.

I presented the orders I prepared to Major Shibabaw.

"Keep it with you. You are the one who is going to present the briefing." I did not like this, but there was nothing I could do. We got into the provided vehicle and left for division headquarters.

The officers were in the briefing room at 0745 hours to listen to the battle plans for the upcoming operation. We entered the briefing room and took our seats. The Deputy Commander took a roll call to make sure all those required were present.

A while later, the 2nd Division Commander arrived. We all stood, saluted, and resumed our seats.

The Commander called upon Major Shibabaw to present the battle plans.

Major Shibabaw stood up and said, "Sir, the report will be presented by Captain Tesfaye." He then took his seat.

"That is all right," the division commander replied. Turning to me, he asked, "Captain Tesfaye, are you ready to give us the operation order?"

"Yes, Sir!" I replied. Standing, I saluted and asked if he was ready for me to begin.

"Go ahead," he ordered.

Addressing the assembled officers as though they were the Expeditionary force I began to present the operation order which I had prepared for my troops the day before.

Much of the initial information in the battle plans was merely a review of previous information concerning the current situation at Nakfa, estimated enemy strength and capabilities, the dire situation the 15th Battalion was in, as well as local topography, communication plans, etc. This information had all been repeatedly covered, so I moved quickly through it. Soon, I was discussing the battle plans I had developed.

Mission

One company consisting of ninety-nine soldiers of the Airborne Battalion's Expeditionary Force will fly from Asmara Airport in two C-119 airplanes made available by the Air Force. The soldiers would parachute in over Nakfa and reinforce the 15th Battalion as ordered.

Execution

The C-119s can each transport fifty soldiers. Therefore, the company including the headquarters element is divided into two platoons for the duration of this mission. The two C-119 airplanes will be loaded for the jump with the first platoon in the first airplane, and the second platoon in the second.

Before our jump, a pathfinder will insert into the 15th Battalion's encampment from a helicopter to mark the drop zone at the proper time.

Because the mountain range is on the western side of the area held by the 15th Battalion, the best flight direction is north to south. The length of the territory controlled by the 15th Battalion, from north to south, is no more than 1000 meters. The first jumper, to avoid missing the drop zone, will jump 100 meters inside the starting point of the dropping zone while the last jump will be made 100 meters short of the end of the drop zone.

We need to keep all jumpers within the 1,000-meter drop zone. Therefore, 200 meters will be deducted from the total length of the drop zone, effectively leaving us with an 800-meter drop zone. An interval of 100 meters will be maintained between each jumper to avoid collisions or entanglements while in the air.

During the drops, the aircraft will be in danger of taking fire from enemy anti-aircraft batteries. Therefore, we will reduce the time that the plane is over the drop zone as much as possible. To achieve this reduction in time, we will shorten the intervals between jumps. Three soldiers will jump every two seconds. This rate will allow us to complete the drop in just two passes.

Because the drop zone is such a high-risk area, the first group must jump on the first pass. The jump-master will have the jumpers ready when the pilot adjusts the direction towards the white stone houses. The panel markers may not be visible due to the high altitude of the flight or may not even be out due to enemy action. Therefore, the jump-masters will be prepared to make the jump by using the white stone house with the tin roof as a point of reference.

Equipment inspections will be completed while at the airstrip. We will await the order of the jump-master before boarding the planes by assigned jump groups.

During the jump, soldiers will direct their descent towards the T-shaped panel. If they cannot spot the T-shaped panel from the air, they will direct their descent toward the two houses made of corrugated iron. The two corrugated iron houses will be our assembling point.

As soon as we are on the ground, we will free ourselves from the parachute, pull our weapons from our bags, and set up a defensive perimeter. We will defend our position and provide cover fire for those still making the jump.

As soon as we have our position control, we will establish contact with the other team and then with the 15th Battalion.

If a jumper lands in enemy territory, we will provide cover fire if possible and that soldier should make every effort to rejoin our company.

Each soldier will have their rifle, a bayonet, 360 rounds of ammunition, five magazines, and four hand grenades. Those soldiers operating crew-served weapons will also be issued .45 caliber pistols and two loaded magazines. M60 machine gunners will carry three regular combat loads of ammunition. Each squad will have one M60 light machine gun, and one M69 grenade launcher. Each soldier will carry a full combat load.

My briefing went on to cover the rest of the equipment, parachutes, communications, and other crucial elements of the operation. I ended my presentation with a discussion of command and control.

Command and control

I would be in the first plane together with the first platoon and would be the first person to jump. After landing, I would remain in the assembly area until all soldiers have assembled. During the mission, I would be with the first platoon initially. But, as the action unfolds, I could move to wherever I need to be to best command for this operation.

I went on to explain that if I were killed or wounded to the degree that I could not remain in command, the first platoon leader would take over company command in addition to the command of his platoon.

Finished with my briefing, I announced I was ready to answer any questions.

"Thank you, Captain Tesfaye," said the Commander of the 2nd Division. "If there are questions, they will be asked later. For now, we will proceed directly to hear the operational plan for the Air Force." He, then, invited the air force to present its report.

The operation officer of the Asmara base Air Force delivered his plan as follows:

Operation Plan of the Air Force

The Air Force has prepared two C-119 aircraft, each with the capacity of carrying 50 soldiers and their equipment for the jump into Nakfa.

The Air Force has tasked a squadron of F-5 fighter jets to provide air cover for the operation. Sufficient fuel and munitions for the jet fighters are already in readiness. Maintenance teams will be standing by with the necessary spare parts to expedite any needed repairs.

Emergency crews will be on standby at the airport to handle ay flight emergencies, and emergency rescue teams will be placed on high alert together to rescue any downed pilots.

We have assigned a helicopter for the Pathfinder's jump. The Pathfinder's flight will also be protected by a fighter jet during his insertion.

One squadron of fighter jets, organized into two teams, will provide continuous air cover over the Nakfa area. The C-119 airplane will be protected by fighter jets during the flight from Asmara airport to the Nakfa area.

While one flight is refueled and rearmed at Asmara Airport, the other flight will fly combat patterns over the region track movements of enemy infantry and anti-aircraft gun positions. When they spot a target, it will be taken out.

After the jump, continuous air support will be provided for the Airborne unit and the 15th Battalion, as they join forces and attack the enemy.

This presentation completed the Air Force briefing.

The 2nd Division Commander invited anyone present to make comments or ask questions.

After a brief discussion session, the briefing was adjourned, and we returned to our military camp. We had a busy time ahead of us.

CHAPTER 9
The Parachute Jump at Nakfa

THE TROOPS COMPLETED their preparations and began rejoicing. Some played volleyball while others sat thinking and smoking cigarettes. On October 30th at about 1500 hours, it was time for Sergeant Geremew Haile Mariam, our Pathfinder, to leave for his mission. A few of us left for the Air Force base to see him off.

As the helicopter pilot and his flight team waited, we said our goodbyes to Sergeant Geremew, wishing him well as he left.

The helicopter lifted off with Sergeant Geremew and headed north-west, flying up into the clouds. Ten minutes later, two F-5 fighter jets took off. Major Shibabaw and I went into the Air Force operations office to follow up on the situation.

After 45 minutes, we were told that the helicopter had reached the Nakfa area. I feared that the helicopter might be hit. Unable to sit down still, I paced about the office anxiously. Five minutes later, we were told that the Pathfinder had safely reached the position occupied by the 15th Battalion. Those of us present in the office embraced each other, and, although we could not thank God openly, inwardly, I was certain we all thanked God. We returned to our camp. Soon, we received a telegrammed message of congratulations from the 2nd Division.

When the troops heard the good news that Sergeant Geremew had landed safely and joined the 15th Battalion, they were ecstatic. Sergeant Geremew is a specialist as a Pathfinder. A Pathfinder is a man who jumps into a drop zone first to put up markers for wind speed and direction and to mark the beginning and end of the drop zone.

The next jump was scheduled to take place on Oct. 31st at 0700 hours. We finished our preparations and spent the evening talking with the troops. We went to bed early because we had to get up at 0200 hours. However, it was very difficult to fall asleep. I began to think of my pregnant wife and our four little children and wondered what their fate would be. Without having closed my eyes even for one minute, the Sergeant Major blew his whistle. It was time to get up. I was lying there in my clothes, so I got up and immediately headed to the parade-ground where the troops would assemble.

The troops quietly assembled in ten minutes. Orders were given to complete preparations quietly and soldiers were assigned to trucks for the ride to the airstrip. In accordance with these instructions, our men loaded up and, carrying their parachutes, climbed into the vehicles assigned to them and took their seats. Once everything was ready, I gave the order to move out.

The convoy of vehicles rolled to the airstrip where the men, gathered up as quickly as possible all their equipment and weapons and dangling parachutes, climbed out and formed up on the tarmac. Soldiers not making the jump were present to help speed up the preparations. The area where we readied ourselves was well lit which helped. Sergeant Negatu Tegegne, who oversaw the jump on this day, took charge and giving us clear instructions, got us prepared to board the plane. I lined up in my group with the other soldiers. Following Sergeant Tegegne's instructions, we moved quietly.

Great care was taken with the parachutes to ensure there would be no chute failures during the jump. Each man would carefully lift the parachutes off the ground and help secure them onto the back of the man in front of him. When the first man was certain that his parachute was secure, he turned to help the jumper behind him. Following this process, all the jumpers donned their parachutes.

When a jumper stands before a jump master for inspection, the jumper puts his two hands on his helmet and stands at attention. If the jump master has a correction to make, he moves the jumper one step forward or backward out from his row. The jumper then corrects the error under the watchful eye of the jump master, and then waits until the jump master checks to ensure the error is corrected. Once the jump master is satisfied, he returns to his place in the line.

Jump masters are very strict and thorough. A small error may claim a life, so no error is overlooked or tolerated. The inspection process was finished at 0500 hours. It is not easy to sit or stand for a whole hour wearing a parachute and holding all your equipment. The weight of our parachute and equipment made it hard to sit or stand, so a jumper will seek support from another to hold his hand and support him until he lowers himself to the ground. The pilots warmed up their aircraft's engines and flight crews completed loading the fighter jets with munitions. Flight technicians carried out the required pre-flight inspections.

Asmara is very cold in October. The cold penetrates deep into one's bones. The wind makes it even colder. Even in the dawn's sunlight, the cold makes the teeth clatter, stiffens the skin, and makes the hair stand. One hour seemed to last an eternity. Everyone checked their watches constantly, wishing for the time to move more quickly so they could escape the biting cold.

Finally, though it seemed like an eternity, 0700 hours arrived. A few minutes later an Air Force officer arrived and announced that they'd received a radio message from the Commander of the 15th Battalion, that the Nakfa area is covered with fog. We could not take off until they tell us that the area is free of fog.

The troops grumbled. They were bitterly disappointed at this delay. As there was no other alternative, they began to pray that the fog would clear soon. As if we were not frustrated enough, two hours passed by with no change in the weather. Three, four, five, six hours passed by. Still, no change in the weather. The parachutes bent our spines backward and we were struggling to keep from hurting our backs. We could not take off the parachutes as it would require too much time to put them on again. While we were baffled by the fog that covered Nakfa, the humid wind blowing from the Red Sea suddenly came to cover the Asmara airport with fog as white and as fluffy as cotton. Briefly, the fog disappeared for a few minutes, but then came back suddenly to cover the airport once again. The fog covered the airport until 1500 hours.

At 1500 hours we were informed that the scheduled jump for the day was cancelled and to come back the next day at 0300 hours. We helped each other remove our parachutes and putting them into bags, returned to our camp in the same way we had come. We wished the jump was over with and we wondered how many of us would now be dead if we had jumped. It was good to be alive even for one more day; but as for tomorrow only God knew.

After we arrived at the camp we unloaded our equipment from the vehicles, secured our weapons and parachutes in the same order as we had done previously and went to the dining hall. We were served better food than usual. After dinner many soldiers, who were completely exhausted, went to bed and quickly fell to sleep. There was no noise in the surrounding areas. Other soldiers seemed to be sleeping, but they were wide awake and thinking of their families.

At exactly 0300 hours, the whistle sounded. The troops quickly ate their meal, drank their tea, and lined up for roll call.

Quietly and methodically, our company went through the same preparation steps we had completed the previous day, readying ourselves to make the jump into Nakfa to relieve the 15th Battalion. There were a few clouds but nothing that appeared problematic enough to hinder our aircraft from taking off. The weather around Asmara had improved since the previous day. We again sat, waiting strapped into our parachutes, weapons and equipment, and hoped the operation would get underway.

At 0700 hours, an officer arrived from 2nd Division operations.

"Where is Captain Tesfaye?" he asked.

"I am right here." Raising my hand, I responded

The officer handed me a telegram message from Sergeant Geremew. The message stated that due to the intensity of enemy fire, it has not been possible to map out the drop zone. As a starting point, try to use the 15th Battalion camp and the residence of the former provincial governor. It was the only stone house with a tin roof and should be easy to spot.

I informed my Battalion Commander of the situation. The 2nd Division Commander and the Asmara Air Force Base Commander were conversing nearby, so he quickly went over to them and updated them on the situation. I assembled the jump masters and updated them. The pilots and jump masters moved off to further discuss the situation. Shortly thereafter, we received another message from the 15th Battalion recommending us to wait patiently. The weather had not shown any improvement yet.

Then at 0930 hours the order came to get ready to board the planes. I ordered the troops to line up at once and reported to the 2nd Division Commander that we were ready to go. He saluted us and wished us good luck. I ordered the jump masters to have the troops board the aircraft. In a few minutes, we were on board and ready to take off. The fighter jets started moving down the runway in twos, following each other. The first two fighter jets were already moving in line for takeoff. The next two fighters followed close behind, to take off soon after. The remaining fighter jets were lined up on the taxi way ready to immediately take their turn. Soon we were airborne.

The jump master stood in front of the troops, "Airborne! Airborne!" The troops were nervous and excited. The jump master shouted again, "Are you going to jump?"

The troops answer in unison, "Yes."

"When?"

"Now."

"How?" said the dispatchers.

"Like a tiger," answered the troops.

"What are you?"

"Flying tigers!"

As the aircraft approached the drop zone, the jump master ordered, "Stand up!"

The soldiers in the first stick rose to their feet and clipped the snap hooks on their static line to the anchor line. The jump masters then checked to make sure these were properly secured. Next, they checked that all static lines were not intertwined. The jumpers collected and held their remaining static-line to keep it clear.

I would be the first jumper in the first team. Thirty minutes passed since we left from Asmara. To avoid detection by the enemy, we flew southward in the direction of Barentu and Tessene. After 40 minutes, the pilot banked the aircraft, changing the flight direction from west to north. All eyes were on the red, yellow, and green lights at the front of the aircraft.

To alert us that we are approaching the drop zone, the pilot sounded a bell and turned on the red light. When the light changed to yellow, I stepped forward and stood in the door.

The jump master standing at my side was Sergeant Negatu Tegegne. I awaited the jump master's order to go.

The wind whipped past the aircraft's door. When I tried to look out of the aircraft door the wind twisted my face to one side and tears filled my eyes. I could barely open my eyes, but I could make out in the distance a chain of mountains and a dry riverbed. The town, however, was not visible. Something like a white cloth flat on the ground came into view. I assumed it was the building with the white tin roof.

"Do you see the white building, sir?" Sergeant Tegegne asked.

"Yes," I said to him.

The green light flashed, and I heard Sergeant Tegegne shout, "Go!"

I turned to my men and shouted, "Follow me!" Then I jumped through the door. The first batch followed me out the door.

The altitude we jumped from was so high that all the mountains looked like they'd been flattened into the ground. The houses and the trees were indistinguishable from the ground, and all below looked like a flower carpet with many decorative designs.

As my parachute slowed my descent, I looked around me. Due to the speed of the wind, the high altitude from which we had jumped, and the differences in the abilities of each jumper, it caused us to scatter as we descended toward the ground. We all fought with the suspension lines of the parachutes to land closer to the white house. The man who jumped immediately after me was my radio operator. I looked for him in the air and discovered we were a good distance apart.

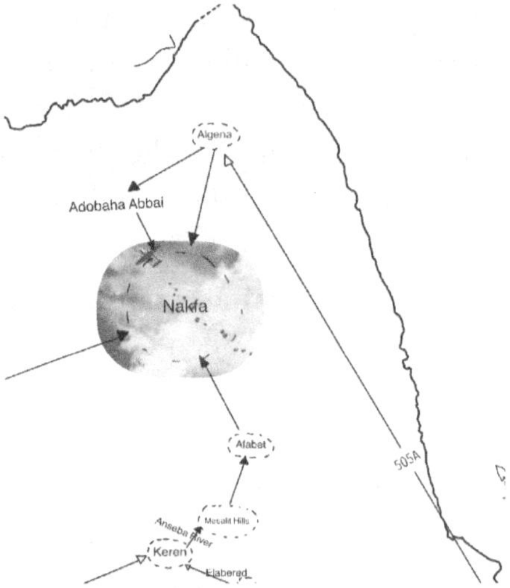

Parachute jumping mission, 1976.

We all worked to narrow the distance between us by working the lines of our parachutes. I looked and could not see many other jumpers in the air. I began to search for the two aircraft, but they were nowhere to be seen.

I could see the F-15 fighter jets working to suppress any enemy action. They looked like eagles, diving at a fantastic speed towards the ground, wreaking havoc with their weapons, and climbing once again into the sky. I thought that perhaps the two transport airplanes were hidden behind the chain of mountains to the west. I looked intensely in that direction, but I could see nothing.

I thought to my great surprise, that I heard a waterfall, but that would be impossible, especially in the area we were jumping into. I quickly discovered, however, that what appeared to be sounds of a waterfall turned out to be the sounds of bullets whizzing past me in the air. The enemy was shooting at us. I was terrified as I heard rifles, mortars and launchers exploding around me. I struggled with the parachute suspension lines. My head was filled with a jumble of thoughts. Where were the two airplanes carrying the rest of my men. How far did the men land from the white building? How many of them landed in enemy territory? And, where would I land myself? I looked carefully for the white building.

I realized that I had drifted far to the east of the white house. At this point, the distance between me and the ground was no more than 300 or 400 meters.

To avoid landing in enemy territory, I used all my strength to pull the right suspension lines of the parachute, and to make the canopy on the western side fold inwards. By raising the canopy on the eastern side upwards, I increased the wind push towards the west. I followed the drop of my radio operator with my eyes. He was quite a distance from me. I knew that if I had no radio, even if I made it to the ground, it would be very difficult for me to coordinate the activities of the troops.

The bullets began to penetrate the parachute. The enemy knew that the first jumper commanded the troops, so the enemy was purposefully targeting my parachute.

I looked to see that I was over the drop zone. Thankfully, I was not over enemy territory. I pulled at my parachute's suspension lines, guiding my descent as best I could as I approached the white-roofed building.

I could see enemy soldiers on the ground beneath me. I knew they could not be soldiers from the 15th Battalion because we had been informed in Asmara that the troops of the 15th Battalion were taking heavy enemy fire and would not be able to leave their trenches to support us. Despite the fighter jets flying over them, the enemy was still moving to engage us.

When a jet passed overhead, they ducked and scattered. Two of what I believed were enemy soldiers ran towards me and I only had about 50 meters between me and the ground.

The ground underneath me was covered with rocks. My training took over as I prepared to land and hit the ground in a good parachute landing fall. On the ground, I heard men shouting but I could not make out what they were saying. I guessed that these were enemy troops shouting at me to give up. I assumed they were coming to capture me.

Quickly, I disentangled myself from the parachute, retrieved my weapon from its bag and putting my finger on the trigger, prepared to engage the enemy.

"Don't shoot! Sir! We are soldiers of the 15th Battalion! We left our trenches to help you when we saw you jumping down to die with us."

"Don't move!" I replied. "Give me proof that you are indeed members of the 15th Battalion!" The men stopped moving toward me.

"One of your men, Sergeant Geremew, who jumped earlier this day tried to spread out panels indicating the drop zone. He was prevented from doing so by enemy fire. Right now, he is with the battalion commander and is helping direct the strikes of the fighter jets."

As they were telling me this, a fighter jet zoomed past over our heads without firing on us. This was the proof I needed that these men were on our side.

"Where are the other men and what happened to the two aircraft?" I asked them. There was some confusion among them

"Oh! They have been hit by the enemy's anti-aircraft weapon, and they returned to Asmara," said one of them.

"No," said the other, "they returned after completing their mission."

Indicating the direction with my hand I asked them to go and get my radioman who landed a little to the north of where I landed.

"We were given instruction to help you," they replied. "We cannot leave you alone. If you agree, we can go with you to find him." After we had gone a short distance my radio operator, accompanied by some members of the 15th Battalion, came running towards me. We stopped and waited as he approached. He had also landed safely.

"Establish communications with our platoon leaders and the Battalion Commander." Soon he had the first platoon leader and the battalion commander on the radio. He handed me the handset of the radio he carried on his back.

My call sign was Tiger. The 1st platoon leader was Tiger One and the second platoon leader was Tiger Two. The commander of the 15th Battalion was called Lion. Taking the handset from the radio operator, I addressed my platoon leaders.

"Tiger One and Tiger Two, are you there." They both responded affirmatively. I told them to pop smoke to indicate their locations. Tiger One used a green smoke grenade. Tiger Two used a red smoke grenade. They did as instructed and judging by the smoke, they were about 500 meters apart.

Because the airstrip was to the north of Tiger One's position, I told him to hold his position. I ordered Tiger Two to maneuver to join Tiger one. I told Tiger One I was moving to his position as well. Once we joined forces, the airstrip would be 5 kilometers to the north of where we were. Our main objective was to take and hold the airstrip to bring in troops and logistical support. The enemy's objective, of course, was to hold on to it. They were currently using it to resupply their positions from somewhere to the north.

I reached the first platoon's position and for a situation report. The platoon leader reported that the enemy held a strong position at the base of the small hills to our front. The enemy was well dug in on both the right and left side of the hills with a well-positioned machine gun as well as mortars and other heavy weapons. He was taking continuous heavy fire from their position.

"They are keeping us pinned down. We can't even raise our heads. We have to crawl like worms!"

Our fighter jets had tried to destroy the heavy weapons hidden in the valleys. As the weapons were well-hidden in the valleys and hollows of the mountains, it was hard to take them out by air. When the fighter jets approached, the enemy stopped firing to keep from giving their position away. When the jet left the area, the enemy immediately resumed firing.

To slow the movement of our forces coming from Afabet to reinforce Nakfa, the enemy dug trenches in the ridges around Nakfa valley to secure the high ground. This gave them the advantage. They'd fortified this high ground as well with anti-tank launchers and 50 caliber heavy-machine guns, and other weapons.

We waited. To get any reinforcements or logistical support, and to remove our wounded, we had to take the airstrip from the enemy. The airstrip was higher ground. Attacking the enemy directly to liberate the airstrip could result in heavy casualties, so we needed another approach.

I got on the radio. "Tiger Two, this is Tiger."

"This is Tiger Two," he replied.

"Give me a situation report." His position was not any better than ours.

Advancing now, with our few numbers would be a waste of time and men. I chose to stay where we were and hold our position while waiting for the second stick of jumpers. I again got on the radio and directed Tiger One and Tiger Two to hold their positions and send out scouts to look for enemy activity. We would secure this position for the send group of jumpers.

Tiger Two now found himself in a very steep area. He came over the radio, "Tiger, this is Tiger Two."

"This is Tiger."

"The enemy is holding higher ground and is inflicting heavy damage on us. I want to move forward and take the high ground from him."

"All right," I said to him. "Hold until I get you cover."

I had the first platoon transfer two machine guns and one rocket launcher to Tiger Two so that they could give him some additional fire power. I got on the radio and had the Pathfinder, Sergeant Geremew, direct our fighters to attack the enemy entrenched on the ridge at the western side of the Nakfa valley.

Coordinating the additional machine guns and launchers transferred from the 1st Platoon with his own two machine guns and two launchers, Tiger Two moved his troops into position and waited for the fighter jets to begin their attack.

Four fighter jets began their strikes, coming in at a very low level from the south. They attacked the enemy positions, one after another, raking the enemy with bullets.

Taking advantage of the cover provided by the fighters, Tiger Two had his machine-gun team move its position forward, and firing from the flank, forced the enemy to take cover. Second platoon moved forward to attack.

Wasting no time and like an arrow, second platoon flew into the trench in which the enemy had taken cover from the jets. Quickly overwhelming the enemy, they soon had control of the defensive position. Immediately pushing forward, they proceeded to the second trench and soon had it under control. Seeing this, many of the enemy were terrified. They began to retreat, abandoning their trench. Taking advantage of the situation, the fighter jets pursued the enemy that were now fleeing across open ground and hit them hard.

In the meantime, the two C-119 airplanes had arrived with the second group of jumpers. I passed an order to the 2nd Platoon to stay where it was, set up a defensive position and assist the jumpers landing in that area.

More fighter jets began dropping bombs on the enemy anti-aircraft weapons, enabling the C-119 planes to fly low. In one pass, 25 jumpers jumped from each plane. The aircraft began their return flights to the airbase, and all our paratroopers landed on the ground held by our forces. Efficiently stripping off their parachutes, they proceeded to the assembly area assigned to them, their weapons at the ready. Quickly, they formed into their squads and joined their platoons.

The second group was given a situation report by their platoon leaders. I sent the new troops to relieve the troops who had just engaged the enemy, ordering them to sweep forward from the point that the enemy had started fleeing and to the east. The second platoon, firing and maneuvering, began our push toward the air strip.

I instructed the Pathfinder, Sergeant Geremew, to coordinate air support for the second platoon as they advanced. The fighter jets immediately started to bomb the enemy trenches. The second platoon pushed forward behind cover fire provided by the jets. First platoon, having now improved its position, informed me that it was now able to begin their own attack.

After a short fight, the enemy abandoned its trenches on the ridge east of Nakfa. Taking the eastern trenches by our forces struck fear in the enemy's heart. However, they had a lot of mortars and anti-aircraft guns which they did not want to abandon. They put up stiff resistance trying to secure the weapons.

I moved some men from the 15th Battalion and had them set up a defensive position to the north, shifting the first platoon to the west. I then ordered an attack to seize the enemy's mortar positions and anti-aircraft guns.

Second platoon completed its mop up of the ridge and continued its assault towards the airstrip. The enemy forces, seeing second platoons advance toward the airstrip, began a strategic retreat, using fire and maneuver tactics to slow second platoon's advance.

After an hour of heavy fighting, we had the airstrip completely under our control. Seeing the direction in which the first platoon was advancing, the enemy troops began to retreat towards the south, following a dry river bed. Sticking close to the foot of the mountain, they attempted to take their mortars and anti-aircraft guns with them.

They attempted to delay the first platoon's advance by placing a small number of troops armed with light weapons on high ground. First platoon, seeing what the enemy intended, made use of cover fire from their machine guns, and rushed to the foot of the mountain, capturing the heavy guns of the enemy. The enemy soldiers, now out flanked by our men, began to retreat along the foot of the mountain.

Our mission was a huge success. We liberated the area surrounding Nakfa from the enemy and resumed air operations for the captured airstrip. Two helicopters, carrying fresh troops, arrived from Keren. The 15th Battalion, having been under siege for a month, had many wounded men in need of proper care. Medical personnel set up a triage team and began to prioritize treatment for the wounded men, who were loaded onto helicopters in order of the seriousness of their wounds and sent to Keren for medical treatment. From Keren, those needing special treatment would be sent by airplane to Asmara.

Now that resupply operations, and proper care for the wounded was under way, we continued our advance to prevent the enemy from digging in or reorganizing and launching a counterattack. We continued our advance on the retreating enemy troops. The 15th

Battalion Commander and I met to discuss plans to continue the battle. We agreed that since dark was approaching, it would solidify our present position. Throughout the night, we worked to fortify our defenses against any possible counterattack.

We decided that my airborne unit should take control of the region to the east and north, while the 15th Battalion would take control of the western and southern region. We sent patrols out one hundred meters from our position to prevent the enemy from carrying out a surprise attack. We also stationed reserve forces in key areas. Radio checks would be made every full hour. Tight security was maintained. This meant no unnecessary movement, no lights, and no loud noises. Platoon leaders gave these instructions to their troops and ensured they complied. I went around to check that my men were in good shape, had food, water, and ammunitions. The orders regarding security were being followed and the men were of good spirit and morale was high.

I proceeded to the area serving as our headquarters, where a radio operator served me a delicious stew and cornbread supplied by the 15th Battalion. As I was very hungry, I quickly ate the food, washing it down with water. The radio operator had also cleared a makeshift bed for me by using his shovel to smooth and level the ground and remove stones. There was even grass for me to lay on. Gratefully, I loosened the equipment around my waste and laid down with my boots still on. I checked my watch and saw that it was 2000 hours.

There was no noise at all. The soldiers who had been through a full day of intense battle, took turns keeping watch and resting. I was concerned, however, that the ones keeping guard may be overcome with fatigue, exposing us to a counterattack by our enemies.

My radio operator, who spent the whole day running here and there to pass and coordinate messages was also very tired. After I had slept a few hours, I took over their radios and allowed them to sleep as well.

While I was going through the activities of the previous day in my head, the radio came to life.

"Tiger, this is Lion."

"This is Tiger."

"Tiger, report to me immediately."

"Understood."

My radio operators had just fallen asleep. Since I had no one else, I woke them up and told them to splash water on their faces. We quickly got ready and then made our way to the 15th Battalion headquarters.

I had not yet met Major Mammo. Major Mammo Temtem was a short, stocky man of middle age, strong and active. Later I learned that he had been transferred as an NCO from the regular army to the bodyguard unit where he had attained the rank of officer. By 1976, he'd attained the rank of major. When I arrived, he greeted me warmly.

"Good evening Captain Tesfaye. Thinking you would come, we had arranged a sleeping place for you. Why didn't you come?"

"I thought it would be better for me to stay with my troops," I replied.

"You could easily coordinate their activities from here," he went on.

"I could, but in Airborne units we prefer for the commander to be with his troops." I thanked him deeply for the preparations made for me and asked him why he wanted me now. He pulled out a piece of paper from his pocket and handed it to me to read.

Congratulatory telegram message regarding the successful jump

From: Commander of the 2nd Division

Asmara.

To: Commander of the Airborne Company

Nakfa.

Congratulations, first of all, for the successful jump. Our entire army expresses its admiration for the decisive step you took and for your heroic accomplishment. We are not able to send the planned relief column as soon as we had hoped. It is going to take some more time. As explained to me by the commander of the 15th Battalion, I understand that your present position is exposed to the heavy guns of the enemy, and that due to a shortage in men you are scattered. I can see from the report that the enemy could easily counterattack over the unoccupied area. Until the planned relief can be completed and implemented, I order you to move your troops from their present position and take up the position that was formerly held by the 15th Battalion. Set up a joint defense together with the 15th Battalion, and Major Mammo will be the overall commander of the combined force.

After reading the message, I was upset. Why send this order without consulting me? How can we let go of territory with such a strategic military advantage and obtained at great sacrifice, and imprison ourselves in trenches? It made no sense. I asked Major Mammo for an explanation.

"This was not my idea," he answered. "I did not give such a recommendation. What I said was that the area under our control now is vast, and that to maintain our positions we need more soldiers, and that reinforcement should be sent quickly." He then called out for his radio operator.

"Sergeant Messel, bring the message you sent to command and show it to Captain Tesfaye." The sergeant showed me the message. Sure enough, the message asked for urgent reinforcement of troops. There was no mention of returning to the previous defensive positions.

I still did not like the order that we should abandon the airstrip, the ridge in the Nakfa valley, and the chain of mountains and pull back to the former position. We now had the advantage. Pulling back to the old trenches meant putting ourselves back in the same untenable position the 15th Battalion had previously been in.

"If high command can send us reinforcements, let it do so. If it cannot, then we must make the best use of our available strength and hold our current position. Can you request command to allow us to do so?" I asked Major Mammo.

"I can't do that," he replied. "If you want, you may prepare the message and we shall pass it on."

I wrote the message out immediately and handed it over to the radio operator to send it. "The Battalion Commander has to sign it to confirm that he permitted the message to be sent," the radio operator said to me. "It is standard army procedure."

I gave the message to Major Mammo who signed it and handed it over to the radio operator to send on.

"I cannot stay here until we get a reply. I must return to my troops. When the reply comes, please send it to me there." I returned to my area. In less than an hour, the reply to my telegram message arrived.

The reply was unambiguous, short and clear.

You are to comply with the order given to you earlier, you have until 0000 hours. Confirm when you have complied with the order.

Now, I had no alternative, but to implement this order.

CHAPTER 10
Life in the Trenches

THINKING THAT THE enemy could be close by, I decided to return to the trenches using the cover of the darkness, rather than attempting it during the day. When it was dark, we left our defensive position and withdrew to the 15th Battalion's location. The battalion commander proposed two options to coordinate our defensive strategies.

The first option was to assign a distinct area to the Airborne company so that we could defend it independently. The second option was for the Airborne troops to join the 15th Battalion so that we could put up a common defense. He asked me to choose between the two alternatives.

I considered the advantages and disadvantages of defending ourselves independently within our own defined area. I decided it was better for us to combine our forces and put up a coordinated defense. No matter how strong one section might be, if the other section falls, the results would be catastrophic.

Also, if we separated from the 15th Battalion, we would have no access to supplies except our water cans. We had no food, or even a kettle for boiling water to make tea. We were in an extremely difficult situation. We needed the 15th Battalion's support to survive. I decided that we should join with the 15th battalion. We simply had no other viable choice.

The defense position of the 15th Battalion at Nakfa had several features. It was established in the southern part of Nakfa town, with the battalion's headquarters at its center. It was established on higher ground that overlooked the neighboring lands, and therefore provided us with good control and oversight of any enemy movement during day light from a short distance. However, a distance to the west, a chain of mountains ran from north to south. An enemy forward observer positioned in these mountains, could direct enemy fire on our defenses.

Running along the base of the mountains there was a wide and deep river bed filled with sand, formed by the flood from the raining season. This river bed provided good cover against our light infantry weapons. This dry river bed enabled the separatists to travel back and forth with relative ease.

Just to the south of Nakfa another shallow dried river bed forked off from the bigger one. This dry river bed ran from our battalion headquarters from east to north.

Along the eastern side of battalion headquarters, about 500 meters away, was the defensive position of the first company of the 15th Battalion. This was not strategically a good position, sitting along the western side of the ridge dominated the defense position of the second company. If the enemy controlled the ridge, they had control over the entire area, including the town of Nakfa. A short distance to the north of battalion headquarters, sat the fourth company position. The fourth company was the 15th Battalion's heavy weapons company.

Theoretically, an infantry company is comprised of 120-150 troops. However, due to the casualties suffered over the last two months, company unit strength for each company was, on average, 50 soldiers. Every day, more men were wounded or killed. Heavy casualties, combined with a lack of reinforcements, was taking a heavy toll.

When a soldier was taken out of action by death or injury, this loss created a gap in the position. Since all our defense zones were of equal importance and under strength, it was not possible to simply move men from one position to another to fill these gaps. Therefore, I assigned my first platoon to join the second company of the 15th Battalion and my second platoon joined the first company. This helped fill the gaps.

During our previous parachute jump and the heavy fighting on that day, we lost only ten percent of our men. One of my soldiers was killed in the air, while a second one died fighting on the ground. Six others were wounded.

On the separatists' side forty-five men died. And, information from Asmara indicated that approximately one hundred and fifty were wounded. We'd captured several light and heavy weapons from the enemy and taken important strategic ground from them.

I was still upset over being ordered to give up the hard-fought ground which had such great strategic value. It was especially hard to give up the airstrip. We had no other way to receive any supplies. To abandon our position and retreat into the trench with the 15th Battalion was bad tactical decision.

It didn't take much time for the enemy to reoccupy the territory we gave back like gift. That same night, the enemy moved back into their positions and began to increase the fortifications.

Airborne units are offensive in nature and do not train to dig trenches and fight protracted defensive battles. After an airborne unit attacks and takes a target, it turns it over to the infantry and prepares itself for the next offensive. On some rare instances, when an airborne unit is to fight a defensive battle, it should only be for a very short period. We do not have the same logistical support as regular infantry units.

The trench dug by the 15th Battalion was dug as a temporary position, and therefore, was not a well dug position. It did not offer protection from heavy weapons fire. Mortar shells caused great damage and inflicted casualties. There were no tunnels connecting one trench to another, so soldiers had to expose themselves to fire if they needed to leave their trench for any reason. There were no protected latrines, so the men had to go into the open to relieve themselves. The enemy was aware of the trench's weakness and the restlessness of the airborne troops. The enemy, aware of our weak position, posted snipers to cover each trench opening. We were targets as we left to relieve ourselves or to fetch water, etc.

Our troops, determined to not allow the enemy to control their every movement, disobeyed orders and moved about anyway. The enemy shot at the troops moving about. The casualties inflicted on my troops within the trenches were worse than the casualties inflicted on us the day before, when we'd made our jump. It was clear to me that if the situation continued like this, within a short period of time, none of my men would be alive or fit for battle.

That evening, I went to each man in the trenches and advised of the importance of following all safety and security orders. Their refusal to do so was merely bravado and not an act of heroism. I told the men to fortify the trenches to the best of their ability, even if they had to use a sharpened stick. The excuse that they lacked shovels could cost them their lives. After talking to my men, I returned to battalion headquarters. It was nearly dawn.

My men took my words seriously and ceased all unnecessary movement. Our casualties were greatly reduced, though not eliminated altogether. Now, most of our injuries were from fragments and debris from incoming mortar rounds, rather than movement during the day. Only one soldier was killed by a sharpshooter.

That evening, I went to the defense position of the second company, to give instructions and encouragement to my men. While I was doing so, all our guarding posts began taking fire from mortars, rocket launchers, machine guns, and other weapons. The firing was intense. I knew it was a prelude to an assault.

I always had a radio with me and I now used it to alert the other positions to be ready. I ordered the men around me not to fire unless they had an enemy in their sights. I told them to lay low in their trenches and not to give away their positions with wasted fire.

"Stay quiet and wait until you see the enemy! Only fire if you are sure to wound or kill the enemy."

A half-moon hung in the clear night sky. All movement up to 50 meters was visible in the bright moonlight. The enemy fire continued without letup. Our men maintained strict discipline and not a single shot was fired towards the enemy.

We waited quietly, listening for the enemy's approach. We realized that their barrage was to disorientate us as their main attacking force approached us. We knew that they were not close enough for us to engage in close-quarter combat. We knew to wait for the mortar barrage to stop and for grenades to be thrown. I told my troops to wait until the grenades started exploding.

"Watch for the enemy muzzle flash," I ordered. "They will tell you where he is." Wait for them to be close enough to engage.

After a few minutes, the hand grenades started exploding. The enemy charged towards our trenches, firing their weapons.

"Open fire!" I yelled. Our men opened fire on the advancing enemy. The enemy troops fell to the ground like leaves from a tree. I could hear the enemy commanders yelling orders.

"Be brave, go into the trench!" Not one single enemy fighter managed to enter our trenches.

The first enemy wave was turned back by our men who were firing calmly and with deadly accuracy. Many separatist soldiers fell dead. The wounded retreated leaving a trail of blood in their wake. A second wave followed with the same results.

The separatist forces didn't like to accept defeat. Our enemy was arrogant. They attacked repeatedly. The result was a great loss of life and a defeat that the separatists had never experienced in their fighting history.

At dawn, fighter jets came to give us support. As their losses increased, the separatists were forced to retreat, leaving behind their dead and wounded. Our second encounter with the enemy ended. Once it was daylight I could not return to headquarters, so I remained with the second platoon. We spent the day repairing and strengthening our position.

Enemy snipers were firing their weapons at moving shadows. We had no respite from their mortar fire which continued all day. The fifty-caliber machine fired on our position repeatedly. A fifty-caliber machine gun is a devastating weapon. The sound is dry and terrifying.

When the fighter jets flew overhead, there was no sound except the enemy anti-aircraft guns firing from their concealed positions in the valley. The separatists knew very well how long a fighter jet could remain overhead before exhausting the fuel in its tank. They knew the ground fire tactics the fighters would use. After the fighter jets made their final strikes and headed back to refuel and rearm, the separatist snipers, and their mortar teams simply came out of hiding and began firing on our position again. We understood their message.

"Your fighter jet wasted its bombs and bullets without doing any harm to us. See, we are still here!"

After what seemed an eternity, the surrounding area became calm. On our side movement was possible only in the evening when it was dark. My radio operator and I started our journey to our own trench once darkness fell.

The men in our various defensive positions went about the work of recovering from the assault. Some, carrying water containers, headed in teams to wells to draw water. Others could be seen digging graves. The wounded, either with the help of crutches or fellow soldiers, headed to the medical center at battalion headquarters. It often took four or five men to carry a wounded soldier. Once it was dark, platoon leaders could get the food from headquarters for their men. Everything had to be done under the cover of darkness.

The only long-range radio was at battalion headquarters. Radio operators were not supposed to communicate outside their assigned communication zones. Sometimes, they went against the rules and exchanged information with other radio operators throughout the country. They were a source of information, both good and bad.

Some of our soldiers who were friendly with the long-range radio operator, would find excuses to sneak off to headquarters and talk to the radio operator. They were interested in whatever information he gathered that day. When one considered the number of troops wandering around outside our trenches, it seemed no one was in the trenches. If the enemy knew about our undisciplined and disorderly movements, it would have taken advantage of the situation and attacked, possibly breaking into the trenches and eliminating all of us.

In addition to my role as the commander of my company, I also had to assist as the battalion commander's operations officer. I told the battalion commander about the risks of the uncontrolled troop movements in the evening and suggested that we find an immediate solution to it.

"What do you propose in the way of a solution?" he asked.

"I will bring you a detailed plan tomorrow morning," I said to him.

"Fine," he said.

Based on what I had noticed the previous night and what I had observed in recent days, I wrote down the following order:

Our forces should not be distracted. Failure to secure our immediate surroundings is a danger to us all.

Without knowing the strength and location of our enemy, troops will restrain themselves from any acts that give away our positions, strength, and available firepower. Any noise we create provides acoustical cover to our enemy, allowing them to move men and vehicles closer to our position.

Contiguous areas between two defensive positions will be secured jointly by the two parties to prevent infiltration by enemy forces.

We will give special attention to those sectors that provide favorable approaches to the enemy. Immediate stop to all unnecessary movements that give away our position to the enemy. Defensive positions will be camouflaged as much as possible.

Having written this order, I gave it to the battalion commander, which was then passed on to the troops.

The separatists brought in two additional brigade-sized elements, each composed of more than 1500 men. They brought in additional heavy weapons, moving men and the heavy weapons to the Nakfa front. Three or four times a day, it rained mortar and heavy weapons fire on our positions. About 250 of our men maintained their positions in the trenches and engaged in a defense battle with M-1 and M-14 rifles. Our men fought not only the enemy soldiers, but also hunger, heat, cold, lack of medical care, psychological warfare by the enemy, administrative obstacles, fifth columnists, and cowards.

CHAPTER 11
Trench Reinforcement

In war, a properly fortified position is critical to a defending force. The trench of the 15th Battalion was not strong and left us exposed to incoming mortar fire. Too many men died. It was clear that we needed to construct better trenches as soon as possible. We decided that our trench should be roofed with strong lumber. We placed sacks, paper or leaves, or any material that could prevent soil from entering the trench over the lumber. Flat stones were then spread over the soil to help deflect the impact of shells landing on the trench to prevent splinters and other injuries to our troops.

However, there were not enough shovels or hoes to dig trenches. Also, we were unable to freely move about to look for the necessary local materials such as lumber, grass and leaves, without the risk of being shot by separatist snipers. The area under our control did not have most of the materials needed for constructing good defensive positions. Whenever we tried leaving the trenches to scavenge whatever material we might find in the surrounding areas, enemy snipers shot at us. It seemed like they had a sniper for every trench.

Despite the dangers and lack of materials, we used whatever we could get to improve our defenses. We shared our tools and worked day and night to fortify our trench as rapidly as we could. After sundown, I would go from trench to trench to check the work and give instructions on proper fortifications.

A trench should have enough width and depth to enable the user to at least stretch his legs. On the other hand, a trench can not be too deep and should enable the user to observe the movement of the enemy and to effectively use his weapon with a clean field of fire. Trenches should interconnect to allow the flow of communication and should also blend in with the terrain. The trench should have a place where members of troops may rest, and where the wounded may recover. The cooking area should be outside the main trench and at a good distance from it.

Some of the troops worked at improving our trench while others collected the necessary materials at night. Working hard, they constructed several, strong trenches in a short period of time. The new trenches provided us much better protection from mortar shells fired by the separatists and the number of deaths and casualties from incoming mortar rounds was significantly reduced after we fortified our trench. The troops were pleased with the results of their work and the old saying that "sweat spares blood" was proven by their actions. The troops from the 15th Battalion, seeing the benefits of a well-built trench, began to reinforce the trenches in their own areas.

The troops now abided by the rules and regulations governing the movements within our defensive positions. All unnecessary movements during the day were curbed. While we still suffered the noise of the separatists' heavy guns and machine guns, the number of those killed or wounded was reduced greatly, thanks to our new trenches.

Due to lack of shelters, everybody stayed in the trench day and night. The trench was narrow and there was no way for the troops to sleep with their feet stretched out. We built shelters, and our men were now able to rotate between the trenches and sleeping in shelters where they could stretch their legs out. Still, they slept without loosening their equipment and with weapons at their sides.

During the six months that we stayed in Nakfa, we never loosened our shoelaces or took off our clothes or loosened our equipment from our waists. We were on guard all the times, day or night.

Between being able to stretch out our legs and resting in comfort and repeatedly driving back the enemy's attacks despite his numerical superiority, our confidence grew. Living in the trenches, killing and dying, became routine. In the evenings, no one would remain in the shelters. Everybody returned to the trenches and pointed the muzzle of his rifle towards the enemy while keeping a sharp eye out for any sign of movement. At night, visibility was poor, so we kept our ears working as well. We remained completely silent to detect the enemy. The men were forbidden to do anything that generated noise or light. Absolute quiet was a tool for survival.

Each soldier was assigned a section of the trench as his position. When the enemy came into range, our men quietly waited, shooting only when confident of scoring a hit. In a single night, the separatists would fake three or four attacks. We were not fooled. The main objective of such attacks was to get us to expend our ammunition. The enemy knew we had a shortage of ammunition. They also used these tactics to try and locate the position of our weapons, particularly our machine guns and rocket launchers.

We pledged not to pull the trigger unless we were sure to kill or wound the enemy. We wouldn't fire back whenever they did. We waited patiently for the enemy to make a real attack. The separatists faked attacks again and again, until dawn. When day broke, both sides returned to their respective roles. They shot out bullets on us from heavy and light weapons and we worked to improve and strengthen our trenches.

Coffee addicts roasted wheat in the recovery area, until it became charcoal black. Then, they ground it in their helmets to make coffee. Cigarette smokers rolled tea leaves in a piece of newspaper to satisfy their smoking addiction. Maize was pounded in a metal helmet with a stone. It was kneaded in a metal helmet then baked on any flat metal available.

While in the shelter, a soldier offered me the knock off coffee made of wheat. Another who knew that I was a heavy smoker offered me a "cigarette" which was, in fact, tea leaves rolled in a piece of newspaper. I gladly accepted. The troops did everything in their power to please me. The offerings, the jokes, and the small talk were all efforts to relieve my worries. There were so many incredible stories about our life in the trenches.

Military doctrine teaches that, for an attacking force to assault with a defensive position, it should have two or three times more men than the position they are attacking. Typically, if the assaulting forces' numbers and weapons are superior to that of the defending force by more than two or three folds, the latter will not be able to resist the attack for too long. It must succumb eventually. We seemed to be defying this principle. The number of our men defending Nakfa, was less than 300. We had only small arms. The separatists force attacking Nakfa numbered between 4500 and 5000. This meant that the advantage was 15 to 1 in favor of the enemy.

The enemy had 40 mortars or more in its possession while we had none. We had no anti-aircraft or anti-tank weapons. The enemy, on the other hand, had in its possession Soviet made ZU 23 anti-aircraft weapons, 50 caliber and 37mm anti-aircraft weapons. Our enemy had Anti-tank weapons such as57mm and 75mm recoilless anti-tank guns which they used for demolishing our trenches.

Our troops possessed one, or at the most, two hand grenades each. But for separatist troops hand grenades were as common as bullets. Each fighter was supplied with a satchel of hand grenades. It is difficult to state the numbers accurately, but it can certainly be said the odds were very much in the enemy's favor.

Considering the enemy's overwhelming superiority in firepower, they expected to eliminate our small force with its limited weapons in short order and to take control of Nakfa. Separatists troops launched massive attacks on our troops again and again. However, our troops, because of their discipline and training, were able to withstand these challenges and drive back the attacks.

Since our troops had no replacements for those killed or wounded, our number declined with each passing day. We were initially told that brigade would rescue us, but we heard it had been hit by the enemy at

Messalit and most of the soldiers had dispersed. Even this discouraging news did not make our Nakfa troops flinch from their objective. On the contrary, they were more determined than ever before to carry on until the end.

The separatists, continuously reinforced and resupplied, forced the 15th Battalion to stay into its trench. They placed anti-aircraft weapons all around the airstrip at Nakfa so that transport airplanes could not bring in supplies. The only alternative was to drop the materials and food supply by parachute. The parachute used for dropping supplies cost $3,000 to $4,000 U.S. each. To resupply one battalion-sized force required no less than 20 parachutes per drop. This meant that to supply food and ammunition twice per week, 40 parachutes were needed each week. Continually supplying materials by parachutes for 4 or 5 months was very expensive for the Ethiopian government and its economy. Nonetheless, considering the sacrifices being paid by our troops at Nakfa, the Ethiopian government made it a top priority and continued to supply food and ammunition by parachute.

Certain conditions had to be met to ensure that the materials dropped by parachutes from airplanes reached the troops. Ideally, the area where the plane dropped the supplies by parachute could not be covered by anti-aircraft weapons. The plane would pass over the drop zone at a low altitude and low speed. Instruments indicating the direction and speed of the wind would be in place. The markers indicating the direction of the plane's flight and the right and left boundaries, of the drop zone would also be laid out. Our Pathfinder would communicate with the pilot and take corrective measures if needed.

However, when resupply drops were made to the troops in Nakfa, these conditions could not all fully be met. Because of the enemy's anti-aircraft weapons, the aircraft making the drop was forced to fly at a higher speed, out of reach of the anti-aircraft weapons. Because of the aircraft's high altitude, supplies being dropped were blown off-target and would sometimes land in enemy territory or no man's land. When the supplies landed in no man's land, the effort to retrieve them often lead to more dead or wounded among our men. The separatists did not suffer the shortage of food and ammunition like we did. They easily collected the supplies that landed in their territory, but when the supplies ended up in no-man's land, getting to them always involved a fight. Sometimes, the supplies would remain uncollected overnight.

At first, the separatists would run for cover as the jet fighters came in to cover the drop. Our troops could then run out to collect the supplies. This method worked for a time. However, once the separatists realized our tactics, they began to cover the drop zone from several different camouflaged positions. Hearing the fighter jets overhead, our men ran out to collect the dropped supplies as usual, but the enemy now fired from different directions and sometimes inflicted heavy casualties. It was no longer safe for our men to recover the dropped supplies using fighter jets as cover. We now had to do something else.

We used incendiary rounds to burn the supplies that fell in no-man's land so that separatists could not make use of these materials. However, the enemy destroyed the supplies that landed in our territory by means of heavy machine guns. In the end, after all the burning and the destruction, there was little left.

Bad weather had an impact too. It seemed that aerial resupply did more harm than good when you looked at loss of men versus the amount of supplies we received. The 2nd Division finally stopped dropping supplies by parachutes until improvements could be made.

When we had arrived at Nakfa, there was a large store of maize and wheat brought from central Ethiopia, which was meant to be distributed to the drought stricken Nakfa and Sahel residents. Unfortunately, as we began using this store of grain, we discovered that the stored grain was worm-infested. Despite this, we were happy that we found something that would sustain us for a few more weeks if used sparingly. We roasted or boiled the worm-infested maize. Sometimes we used stones to pound the maize in our helmets, mixed it with water, and spread it out to make a kind of bread. Worm-infested maize or wheat, no matter how it is prepared, tasted bitter and caused diarrhea and vomiting when eaten in large quantities. Nobody died from the diarrhea, but many soldiers became very sick. The men suffering from diarrhea couldn't go out of the trench to relieve themselves because of the risk of getting shot. These men had no choice but to defecate in a helmet while still within the trenches. Only when it got dark could we dispose the feces and get relief from the unbearable stench.

Problem of clothing

The missions of an Airborne unit are usually urgent and short in duration and as such, the troops deploy quickly and not encumbered by carrying a lot of extra items. As a result, apart from what we wore, we had only one spare uniform.

We never removed our uniforms for washing so they looked like the clothing of an oil vendor or that of an auto mechanic. Our uniforms had become tattered and torn. The men tried to mend their uniforms by removing the pockets and using the pockets as patches for holes. These repairs were short lived. When we ran out of pockets, we cut the sleeves from our shirts from the elbow down or cut off pant legs from the knee down. These pieces were sewed over the holes. The increasing number of patches created yet another problem by becoming a breeding and dwelling ground for fleas and lice.

The hungry fleas would come out of the sand and launch a bloody campaign for food. The flea bites were so irritating, we scratched with our nails or with a piece of wood for parts of our bodies we couldn't reach, sometimes scratching until we bled. Some of our men would ask how long were we going to be a playground for lice and fleas? How long were we going to wear patched clothing? Others would reply that we would continue the struggle until the government finds a solution to the problem or until we die.

Medical situation

There was only one medic to care for both my Airborne troops and the men of the 15th Battalion. He had nothing; no medical equipment, no medication, no pain killers, nothing for washing and drying wounds, no aspirin, and no antibiotics.

When one of our soldiers was hit by a mortar round and it became necessary to amputate the limb to save his life, the medic had two options. He could let the broken piece hang and rot, or he could cut off the hanging part with a blunt bayonet blade. He had no scalpels or anesthesia. If it was an open wound, we would pull a nylon strand of thread from the suspension line of a parachute and sew it up. We would wash the wound with salty water and spread salt powder on the wound to make it dry quickly. If no salt were available, we would wash it with urine. If the bones could be set, we would find a flat piece of wood to use as a splint, set the bones as best we could, then use a parachute rope to secure the limb to the wood and immobilize it.

We would remove any guns from the hospital shelter for fear that a wounded man would be tempted to kill himself.

From the hospital shelter came cries of agony. Many of the badly wounded men in the hospital shelter had the same question for me when I went to visit them. They would ask how I could be so cruel and watch them suffer like this? I would rather die. Why don't you order someone to finish me off? Or give me a gun and I will do it myself. I did not have a good answer, but I told them that they would recover. Praising their courage, I told them to be brave and patient, and to endure. I would touch their heads to reassure them, and then I left. When I would make later visits, I would always find some of the wounded men dead. But some others would be showing gradual improvement. I saw all types of ghastly combat wounds in the hospital shelter such as amputated limbs, blinded men, stomach wounds. The wounded men's body would be covered with maggots and the smell of rotting flesh was unbearable. Many men couldn't turn over or sit up because they had lost both of their legs and hands. Many lacked the strength to put food into their mouth. The circumstances compelled all our uninjured soldiers to serve as part-time nurses. The gentle care our men willingly and tenderly gave to their wounded comrades, was a remarkable sight. They prepared food for the wounded and fed them, cleaned them the best they could, turned over those who could not move, and when a wounded man died, they shrouded their bodies and buried them.

When our enemy's previous efforts to attack and eliminate our forces failed, they changed their tactics. The tactics they devised were clever. They knew our force had a shortage of ammunition and if we exhausted the ammunition we had, we could not get more. They tried different means to make us waste our ammo.

Under cover fire from their heavy guns, enemy soldiers would fake an attack, and their commanders would shout orders like attack the right flank or charge the left flank! This was an attempt to trick our men into wasting ammunition by firing at non-existent attackers.

The first day, a few troops fired in response, believing that the simulated attack was a real one. The separatists were pleased that their tactic worked, and we had wasted valuable ammunition. The next day, however, everybody understood that the enemy was carrying out fake attacks, and our men held their fire. When the separatists realized that this tactic was foiled, they devised another one. Our defensive position to the west was on higher ground than the surrounding area, so if the separatists could get to the base of the hill, they could make their way to the summit without being seen by our troops. At night, the separatists approached the hill and setup "decoy soldiers" with military uniforms and helmets. Our men, believing that they saw enemy troops, shot at the decoy soldiers. The helmet dropped off giving the impression the enemy soldier was hit. Deceived, our troops open fire on the decoys, wasting a good bit of our precious ammunition.

They were angry when they discovered they'd been fooled and renewed their vow not to waste a single bullet unless they saw a real enemy soldier in their sights. The separatists then tried one more trick. There were a lot of empty ration cans and bottles littering no-man's land. The enemy would try to make it sound as though they were tripping over this litter out in the open. But this did not work. They then resumed their psychological warfare.

CHAPTER 12
Psychological Warfare

THE SEPARATISTS STUDIED our troops downtime activities. When those assigned to guard electric generators, factories, and water dams slacked, the separatists launched an attack to eliminate them. Later in the evening, the separatists announced over a loudspeaker the names of the men it had recently eliminated. They also listed the weapons and equipment they captured.

We were often subjected to propaganda over the loudspeaker. The separatists would call out to us, urging us to give ourselves, saying, "we will eliminate you just as we did your compatriots. The good news is, if you give yourself up peacefully, we will take good care of you. You will be given the option of pursuing your studies abroad in a country of your choice: Britain, America, Germany, or Italy. Or, if you wish to fight Derg with us, you have that choice too. Are you sacrificing your irreplaceable life for Derg, a government that can't even get you basic food, clothing and medication, and does nothing when you are wounded and let's your body rot and get eaten by worms, a government that can't even collect your body when you die fighting, but leaves your corpse for hyenas and vultures to feast on? How could you be so foolish? Give yourself up today and we will take good care of you!"

We would also hear things like, "there are many soldiers who were once fighting alongside you but came to their senses and gave themselves up in peace, like (here they would mention a name). We have since sent them to America and Britain to live in comfort and pursue their studies."

The separatists carried out this psychological warfare on our troops every evening on a regular basis. Eventually, it got to some of our troops. They did not say it openly, but it was evident that some of the troops were wondering how long we would have to continue living in this manner?

It wasn't easy to combat the effects of such continuous propaganda on our soldiers. But it was important to attend to the problem quickly before the resulting morale problems became too widespread. I went from trench to trench to remind our troops of the oath we took. I reminded my soldiers that we came here to fight and die with the 15th Battalion and not to flee or give up. We had to be careful not to tarnish the heroic moment we'd created when we jumped from such a high altitude, a jump which had no parallel in our military history. I reminded them we'd made a pledge to die alongside our brothers and to keep our pledge to our families, our people, and our country.

I asked the men, "Do you really think that they really have your best interests in their hearts? Do not forget that the separatists now launched attacks on us five times a day. Don't forget they were raining mortar fire and machine gun bullets on us not too long ago. They have prevented us from getting our food and medicine supplies. Do you not remember that they have prevented us from carrying our wounded, your comrades, to the recovery areas, and prevent us from burying our dead? They fenced our supply drop zone with their anti-aircraft guns so that our supplies cannot reach us? Why are the separatists so concerned for us now? Don't be fooled with their fake promises!"

I then went on to say to my men, "we must carry out our mission heroically, the same way we started. Death is inevitable. When it is our time to die, we can't escape it, even if we were to hide in our mother's womb. However, if it is not our time to die, we will survive even the most dangerous situations. We shall not give ourselves up to the separatists under any circumstance. Nakfa shall not fall into the hands of separatists as long as we are alive. History and future generations will condemn us if we break the oaths we made to our brothers who fell honorably in battle before us. We will not die a shameful death. And if we must die, let us make our death a respectable one. Let us die honorably, saving the last bullet for ourselves to make sure we are not taken alive. Let us renew the heroic history of Tewodros and that of Balcha Aba Nefso. Let us maintain our fighting spirit and our patience, and fight with honor until the last man and the last bullet."

The separatists continued their psychological warfare efforts as they increased their number and weapons. They also changed their tactics. Earlier, the separatist trenches were about 400 to 500 meters away from our trenches. Now they constructed new trenches a mere 100 meters away from our positions. As the trenches of our troops and the separatists grew closer together, it became difficult for our fighter jets to differentiate between the two. It was therefore decided to discontinue the use of cover fire by our fighter jets to avoid the risk of hitting our troops with friendly fire.

The fact that the enemy trenches were so close meant we could not drop our guard even for an instant. That would be tantamount to passing a death sentence on ourselves and the people of Nakfa. All our troops understood the seriousness of the situation. There was little need to remind them of this. Everyone knew what was at stake. As the situation worsened, even the wounded in the medical aid area asked to be brought back to the trenches. They decided to die facing their enemy despite their wounds. Every wounded soldier, who could still see, hear and pull a trigger, chose to return to the trenches and take part in the fight.

The strength of the separatists kept increasing due to our troops who were captured or gave themselves up. When a trained military force is forced to retreat from its position, any heavy weapons they leave behind were quickly taken by the enemy so they could not make use of them against us. Therefore, whenever possible, if these weapons and machines could not be saved, they were destroyed.

Unfortunately, some of our soldiers, when they retreated or gave themselves up to the enemy, left behind their weapons and ammunition, doing nothing to keep their weapons from falling to the enemy.

Those who were fighting against separatists and later went to join them, turned the muzzles of their own guns towards us to shoot and kill us. While the separatist side was getting strengthened day by day with the weapons it was capturing from us, our side was getting weaker and weaker. We remained prisoners within our trenches, while the separatist move about freely, their troops fighting in shifts. The separatists were organized and sent their wounded troops to the medical center for treatment while burying their dead with proper ceremony. The separatists' forces had become more like government forces and ours became more like bandits.

Towards the end of 1976, the Derg announced that the Birr, the monetary unit of Ethiopia from the previous government with images of the emperor and his family, were being exchanged for new notes with images of people from different sectors of the population. It was therefore required that everyone hand in their old notes to banks in exchange for the new ones right away. Members of the 15th Battalion collected their old birr notes and gave them to their commander, Major Mammo as he left for Asmara. He returned five days later without their new money.

When Major Mammo returned, he explained what had happened to his men saying, "when I went to the bank to exchange your old money, the bank had run out of the new notes. Then I heard that a major battle had erupted, so I had to leave your money with someone and rush back here. Currently, as you all know, there are no roads open or air transport available to send your new notes out here. You will get your money as soon as the battle abates, and the travel resumes. Let us now focus on pushing back the enemy confronting us. This isn't really the time to talk about money matters while we are faced with this life or death situation." The troops were disappointed with Major Mammo's explanation.

On November 21st, at exactly at 1830 hours, the separatists placed a 75mm Recoilless rifle on a small hill to our south and fired four rounds, one after the other in succession, and demolished the southern wall of the stone building. Major Mammo and I were buried under the resulting debris. While I managed to get free of the debris, Major Mammo was still buried under it. I tried to pull him free, but I couldn't do it alone. Other soldiers in the nearby trench heard the yelling and came to help pull him out of the debris. His arms and legs were broken. We were at a loss as to which part of him to hold, his bones were broken in three or four places but we somehow managed to wrap him up in his own clothes and carry him into the trench.

We used flat pieces of wood and tied them to his broken legs on the right and left sides. His open wound was sewed up by means of a parachute thread and sewing needle. He was not able to turn over without help from others. His men had to help him eat or drink. He was suffering greatly. I think he would have committed suicide if he had a gun in his hand.

While we were still dealing with this horrific situation, a small group of soldiers, claiming to be representatives of the 15th Battalion troops, came to me and requested to speak with me.

"What is the problem?" I asked.

One sergeant, whose name I can't remember, replied, "Sir! Major Mammo stole our money while we are shedding our blood. We collected all our old money and gave it to him so that he could exchange it for us for the new ones when he went to Asmara. He has since been giving us all kinds of excuses why we can't get our money back. If he is willing to return the money to us, he can do so now, or else we are here to place him under our control."

"Do you understand of the condition Major Mammo is in?" I asked.

"Yes, we all know," they replied.

"How then, are you going to take him away and put him under your control? When he gets better and can stand and walk on his feet, we will talk to him together, but now is not the time to do so," I replied.

"No, we will take him away on a stretcher if necessary." They were very adamant.

"Well, as long as I am in command I will not allow you to take

Major Mammo away from this trench while he is in this condition. Go back to your guarding post, immediately?" I ordered them.

"Why do you interfere in matters you don't know? You are exposing yourself to danger." They had just threatened me.

"At present, I am the commander of the 15th Battalion, not Major Mammo. How can you say that I am interfering in the affair of the troops I command?" I asked them angrily.

"Sir, you are new. You don't know the full story. Please stay out of this." they responded.

"Don't ruin your heroic deeds and the high levels of sacrifice you have made here with such a heartless act. Wait until he gets well to settle your issues." I said.

"We can't wait. We need to address this now while he is still alive and breathing."

"Do you really want this trouble between us now? While I am alive, I will not give away a man who is hovering between life and death. If you want to take away Major Mammo in his current situation, you will have to kill me first." I said strongly.

They must have suddenly realized that the Airborne company I came with would not be happy if they tried to kill me.

"All right, for the time being we shall return to our posts, but we will be back soon. In the meantime, please think the matter over and we hope that you will have a different proposal for us." They went away. The situation was temporarily resolved, at least for the time being. Then another group of soldiers came up with yet another absurd demand.

CHAPTER 13
The Absurd Demand to Take over the Radio

A FEW DAYS LATER, a small group of non-commissioned officers (NCO) from the 15th Battalion and my Airborne unit came up to me, armed and ready. One of the senior NCO's from the airborne unit gave me a warm salute and stood at attention. Believing that all was well, I responded to his salute and told him to stand at ease.

Still standing at attention, he said, "Sir! We have been fighting against the separatists for months on end. You haven't told us anything about the state of our country. You tell us that Betis Brigade will arrive at Nakfa at any time. But, what we hear from the separatists through the loudspeaker is that Betis Brigade has not even been able to go past Messalit and that it has been hit hard by the artillery fire and retreated to Keren after suffering heavy damage. We've heard that many of its tanks and artillery have been captured, and that many of its troops have given themselves up. Whom shall we believe? You or the separatists?" There was a pause, then he continued. "As of today, we ask that the long-range communication radio be under our control."

"Where did this absurd idea come from? Who is going to be the commander if the long-range radio is under your control? Me or you?" I asked.

"You, of course, we only want the radio," he replied.

"If I am the commander, how am I going to direct my troops?"

The man stated, "We shall give you the information you require."

"When did I ask you for your help? If you mean to tell me that I am not needed, that you have taken over the command, then you should say so. If you are saying that you should be privy to classified information sent between the commander and the high command, this would be illegal. Such illegal demands from you, especially you NCO's who know very well the rules and regulations of the army is unacceptable. These kinds of demands can only lead to our failure. We should remember our oath to fight to the last man and the last bullet. We shall not give ourselves up to the enemy. We shall renew the history of Tewodros. There can be only one commander in any unit. I will not even consider such an illegal demand. Now, since shots are being fired at us, let's not waste our time on this altercation. Go back to your posts immediately."

They accepted my order and left. They never came back to raise the issue again. I believe they regretted their behavior. However, it was obvious that even if this issue was resolved, they would come up with another totally absurd demand.

CHAPTER 14
Unilateral Ceasefire with More Psychological Warfare

When the separatists got on their loudspeaker to spread anti-Ethiopian government propaganda, they started out by saying, "How are you our dear brothers? Today we have a very important message for you, please listen carefully for a few minutes." They would repeat this again and again through their loudspeaker. Our troops, in turn, would fire bullets toward the loudspeaker so that the sound of the microphone would be suppressed by the sound of the shots.

One day it was unusually silent, and the separatists used this opportunity to carry out their psychological agitation. The speech they made was as follows:

"Our dear brothers, we have no quarrels with you. You have done nothing wrong to us. We don't want to see your suffering. You are not our enemies. We know you are just doing what you are told to do. Your superiors are enemies to both of us."

"Your superiors, residing in Asmara and Addis Ababa with their children, their wives, or their mistresses, are enjoying life eating fresh meat and drinking whisky while you are sacrificing your life with no food, medicine, or even proper clothing. Your government dropped you from the sky like trash into this scorching desert. You are subjected to all this suffering through no fault of your own."

They went on to say, "Even the money you authorized to be deducted from your salary to be given to your children and your families was squandered, those who have relatives are getting help but others with no relatives are homeless and are begging in the streets. The children of your superiors are sent to the best schools in the country and are driven around in luxury cars by assigned chauffeurs. Their wives also have assigned chauffeurs and vehicles to take them shopping, to weddings, funerals, and even on vacations. On the other hand, when one of your comrades get killed, their corpse is left in the battlefield for hyenas and vultures. Without proper medical supplies, when you are wounded, your flesh rots, becomes infested with worms, and falls off your bones. You are not supplied with uniforms and you go about in rags with lice and fleas. How is it that a soldier fighting for his country must kneel down and grind worm-infested maize on a millstone and mix the powder with water to feed himself?

For how long, are you going to bear this kind of suffering? Are you hoping Betis Brigade will come and rescue you from this situation? If you haven't heard the story of the Betis Brigade, we urge you to listen on.

The Betis Brigade could not even rescue itself, let alone rescue you. The force code named Betis Brigade was equipped with tanks, heavy artillery guns, anti-aircraft weapons, anti-tank weapons, heavy machine guns, and supported by air. We destroyed Betis Brigade during a battle fought in between Keren and Messalit. The troops who survived gave themselves up to us together with their weapons. Those who did not give themselves up to us retreated in a shameful manner back to Keren leaving behind their tanks, heavy guns, and other vehicles and weapons.

The name was changed from Betis (Breakout) to Dekis (Sleepy) brigade. Everything was wiped out. How could you be so foolish? Especially you of the Airborne Company, you are educated men and can do better. You are our brothers. We don't want to hear about your tribulations or see your suffering this way. It's for your own sake that we are making such an effort. We don't want to see more of our brothers' blood being spilled.

As of tomorrow morning, for two days, there will be a ceasefire from our side so that you can move about without any fear. If you wish, you can come over to our side and we can meet to have a gentleman's discussion with you."

From that moment, not a single shot was heard for forty-eight hours. When some unruly troopers moved from one guard post to another during the day, no shots were fired at them from the enemy's position. Other troops who noted that no shot was fired at these men, started to move out of the trenches.

When fighter jets arrived, and saw no movement of the enemy, they hovered over the sky and turned back. The next day in the morning, a transport plane dropped supplies in to us by parachutes and returned safely without being shot at. When our troops went out to collect the dropped supplies during the day, the separatists still did not fire a single shot. Troop members who trusted each other began to discuss the separatists' ceasefire and the objectives of separatists' messages. Men who had been stuck in their trenches day and night for five successive months, were delighted that they could come out of their trenches and warm themselves in the sun and move about without any fear. When ordered not to do so, many refused.

Those of us in positions of leadership, realizing the problems that could arise and the rebellion that may ensue, could not tell the troops to stop what they were doing. We could surmise that the discussions held in teams by the troops could have a negative influence on the troops and could undermine their spirit to continue fighting.

As was my daily custom to check on my troops in the evenings with the cover of darkness, I once again began to move from trench to trench, this time with a goal to foil the separatists' propaganda with my own counter propaganda.

I started by asking, "Was there a day that separatists have refrained from doing everything they can to eliminate us, the Nakfa troops, for the last successive five months? Why do you think they are being kind now? Is it not because we beat back the attacks with courage and determination? Aren't these words simply used to attain their own goal? They are now just trying psychological warfare when all their other tactics have failed. They are telling us that bitter is sweet like honey, that stones are bread to be eaten, and that what is white is black? And you believe them? What is most ridiculous about the separatist's message is that they will now send you abroad to pursue your studies free of charge, and if you don't want to pursue your studies, they will find you suitable employment. I don't think there is anyone who does not want to be educated or who wouldn't like to live a life of comfort. But do you believe that it is because the separatists love us more than their own people that they give priority to educating you abroad before they educate their own people?

If the separatists have the good fortune to get us all these things, why don't they even fully cover their own bodies when getting exposed to the scorching heat of the day and the biting cold of the night? Why are they living on dry maize and sorghum, and spending their nights and days lying flat on stones on tops of mountains like lizards? Do you think they love us more than they love life itself or their families? If the separatists claim that they love us more than their own people, are we going to believe it? Then, why did they put us under siege for months? Why do they make us suffer and die from lack of medication? Why do they make us starve by preventing us from collecting the food and other materials dropped for us from the air? We are not fools! Men who cannot even recognize this fact. It is all fake. We can't be fooled by it. We shall continue to struggle to the end."

The separatists tried all kinds of tactics to eliminate us but failed. Their last option became declaring a ceasefire for negotiation. The propaganda of the separatists was nothing new to us. Their offer of a ceasefire and call for negotiation was just more of the same. We had seen it all before.

There is no other option for us besides fighting with courage and honor. If we retreat from our oaths, we shall dishonor our dead brothers who fell fighting heroically. If we surrender to the enemy and allow the separatists stand on the graves of our brothers triumphantly, our brothers will have died in vain.

Nakfa is a famous battlefield to which the international media has given lots of attention. The separatists knew this and will do everything in their power to put Nakfa under their control and gain international recognition. Who is going to benefit from the ceasefire and the separatists' call for negotiation? They thought we couldn't see what they are up to. This shows what they truly thought of us.

The separatists had a good reason why they declared a unilateral ceasefire. Our troops had remained in the trenches for months, facing many hardships. The separatists hoped that our troops would surrender, and that they would gain control of Nakfa without a serious fight.

The propaganda only strengthened our resolve as the men began to understand the enemy's true objective. When separatist saw the propaganda was not working, they increased their use of their heavy weapons and started continuously raining shells on our positions.

On our part, we endured the shelling with determination and a calm spirit. Our troops were determined to defend Nakfa to the end.

CHAPTER 15
The End of the Nakfa Siege

THE SEPARATISTS KNEW that we had a strong bunker and that we were determined to defend Nakfa to the end. They were determined to do whatever was necessary to take control of Nakfa. They moved in additional mortars, anti-tank and anti-aircraft weapons, as well as heavy machine guns and pushed forward, moving to about 100 meters from our trenches. They dug strong trenches that could withstand aerial bombardment and got ready to attack.

The separatists started throwing the hand grenades toward our trenches one after the other. Disregarding the explosions of the hand grenades, artillery, mortars and anti-tank weapons, our troops waited patiently to engage them. When the enemy finally attacked, our soldiers gunned them down, one after the other. The dead and the wounded piled up in front of our trenches, and those who survived turned their heads and retreated. Bringing fresh reserve forces onto the battlefield, the separatist made another assault to take Nakfa. They lost many men to death and injury. Then for some reasons, the separatists ceased their activities for a period.

When the separatists suffered a defeat in battle or when hit too hard, they usually pulled back and examined the strengths and weaknesses seen during the action. They would take corrective actions to correct their weaknesses. They would increase their strengths. They collected detailed information about us as well. They prepared for the next battle. We would not get any resupply or reinforcements, so we had no special preparation to make.

I had any wounded man who could still pull a trigger take a position along our trench. The next battle would be one to be remembered by generations to come, written in pain and blood. We were simply not able to resist a major assault by the enemy with our limited numbers and dwindling supply of ammunition. We all ready needed to sacrifice our lives heroically, as we have been shown by Balcha Aba Nefso and Emperor Tewodros II. The battle would not be long; we had no one behind us, no one in reserve, or to supply us with food or water. We could not leave our trenches to fetch water or prepare food.

Every man would have his canteen filled with water as well as two cans of food and two cans of dried bread shall be given out to each soldier. This food would be placed within easy reach or in a pocket if you still had one. Each man would take a single bullet from their magazine and place it in a separate pocket. This bullet would be used to take our own lives when we were certain that we were going to be captured by the enemy.

As usual, we would attempt to shoot two enemies with one bullet, shooting one through the other. If any of our men were to break their oath and try to surrender to the enemy, we would kill him, as he is no different from the enemy. We also burned all orders, secret documents, and intelligence reports.

If one of our defensive positions or guard posts was destroyed by the enemy, those men who survived would move quickly and reinforce the next closest trench. If the enemy pushes us out of the guard post to the west, those troops would move quickly to the battalions headquarter and regroup there.

To continue fighting, we would take the ammunition from our fallen brothers. Everybody would fix his bayonet to the muzzle of his rifle and be ready for hand-to-hand fighting. All activity in the evenings would stop except fetching water, burying those who sacrificed their lives, and administering medical treatment. We equally distributed our remaining ammunition and hand grenades among ourselves.

If we ran out of bullet, we would wait within the walls of the trenches until our enemies entered and engage them with our bayonets. We would fight determinedly and valiantly to the last man.

If I were to be killed or wounded to the extent that I could no longer lead my troops, 2nd Lieutenant Mekonnen Feysa would assume command. This was my final order.

On the morning of March 22nd in 1977, the Nakfa area was covered with a dense fog due to the humid air coming in from the Red Sea. We could not see any further than five meters. Our jets could not give us air support due to the fog. It seemed that even nature conspired with the enemy.

At 0830 hours, the heavy guns of the enemy opened fire on all our defensive positions. Although the first rounds were fired at many targets, the main fire soon concentrated on the western-most defensive position. This position was the key to control all our other defensive positions. Volume of fire from the heavy guns shook the surrounding area like an earthquake. The grass, the trees, the bushes, and debris flew everywhere. The area was covered with smoke and fog.

The shells landed on every trench and it looked like nobody could possibly survive such a bombardment. The enemy troops fired rocket launchers at the openings of the trenches in an attempt to fire into them. It was clear that the separatists wanted to demolish the trenches and anyone in them. They planned to end the battle with this barrage. The shelling came to an end after three and a half hours. The enemy troops then began their assault.

The separatist did not expect any survivors. They didn't think that any of us could have possibly escaped death or injury from that barrage and be able to aim and pull a trigger.

In the western defensive position, we had only two light machine guns and eight M-14 rifles. The troops stationed in this post were swift and courageous. They never missed a target and caught the hand grenades flung by the enemy in the air and flung them back. Our men did not give an inch. Then, at about 1300 hours, the M-60 light machine gun and Sergeant Kebede Assefa were hit by the enemy with the recoil-less M-57 rifle. The machine gun was put out of action, the gunner and his assistant wounded.

While this was still going on, some soldiers from the Airborne unit tied white cloths on the muzzle of their rifles and waved them. They turned the muzzles of their rifles on us. Taking advantage of this situation, the separatists jumped into our trenches. The battle turned to hand-to-hand combat with bayonets. The intense fighting continued. Many of the enemy were killed. Many of my men lost their lives in Nakfa as they had pledged to do. The few troops who survived, followed orders and headed to headquarters to regroup again. Command and control were reestablished. We would defend the water well by joining the troops in the south.

The enemy, reorganizing itself swiftly, brought a force infantry and heavy guns close to our southern area and our headquarters. The time was 1700 hours. The 2nd Division Commander called me over the radio. On the way from my command post to the place where the radio was stationed, I was hit on my right leg. I was not aware that I had been hit.

The 2nd Division Commander came over the radio, asking for a situation report.

"It is over," I said to him.

"I, too, know that it is over. You are authorized to take whatever steps you think are appropriate," he said to me.

"The only thing I can do at the last minute is take my own life when everything is finished. I will not allow myself to be taken by the enemy. Have no doubt about that." I thought for a minute before going on. "If possible, have our fighter jets drop their bombs on us and the enemy."

"All right, I can do that," he said to me. I got off the radio.

When I finished my conversation with the 2nd Division Commander, the radio operator noticed the blood on my leg.

"Sir, I think you are hit." Pulling my trousers up we looked at my leg. My shoes were covered with thick blood. The radio operator quickly took off his undershirt and tore strips from it. Dampening a strip with water, he first cleaned off the blood from the wound and then tied the strip a little above the wound so that the blood would stop oozing. He then covered the wound with the remaining strip. I thanked him and returned to my command post.

Since the bullet remained in my leg, it became quite painful and gradually prevented me from folding and stretching my leg.

I heard fighter jets. I was happy because if we somehow lacked the courage to take our own lives, the fighter jets would destroy us with the enemy. The area was covered with fog throughout the day. In the afternoon, the weather became even worse.

Due to the fog, the fighter jets could not do anything and flew patterns overhead, waiting for the fog to clear. Eventually, they returned to Asmara. The battle intensified in the southern area around our headquarters. That evening, the fog covered the entire area. Men only ten meters apart could not see each other.

Members of the regional police force, guarding the southern area, hidden behind the fog, defected to the enemy. Many of our surviving soldiers, on the other hand, made their way to our headquarters rallying point. The water well had fallen into enemy hands. It was getting dark. At 1800 hours, our headquarters was assaulted by the enemy. The only area that remained under our control was the guard post to the east. We all headed to the eastern defensive position. As the trench was small it could not accommodate us all. Many troops stood outside the trench. Several of them were struck by bullets while standing outside the trench.

If we must die, why should it be by random bullets while standing here? Why don't we attack and kill the enemy, or die in the attempt? What is there left to us? I pondered this question.

We destroyed our long-range radio so that the enemy would not make use of it. We'd already burned any important documents. We had not enough men to gather or move the wounded among us. The wounded outnumbered the non-wounded. The water well that was so critical to our survival, as well as our trench, had fallen to the enemy.

There was little chance of regrouping ourselves to launch a strong enough counter-offensive to recover our trench.

I gathered my troops and asked them what they thought about my plan to attack the enemy. They agreed unanimously that we should attack. Placing our machine guns in the middle of our formation, we lined up and advanced towards the north-east where our enemies were stationed. As we approached the trenches of the enemy, we opened fire. Bayonets on the ready, we rushed into the enemy trenches. The separatist troops fled. We were astounded. We had control of an area free of enemy soldiers.

We headed towards Afabet without waiting for orders to do so. We joined up with Major Mammo Temitime who told us he could no longer move forward. Handing us a letter he'd written to a soldier he trusted, he told the soldier to give this letter to his family if the soldier survived. Then Major Mammo turned off the trail and limped into the rocks and brush A few minutes later we heard a pistol shot. We presume that he took his own life with his pistol.

After walking a while longer with a few tired, wounded soldiers, I found that bullet in my leg was causing me a great deal of trouble and slowing me down. I was carrying one hand grenade, and an M-14 rifle with two twenty-round magazines.

"I will come along slowly as best I can." I said to the soldiers who were accompanying me. "Do not let me slow you down. Keep on going forward." Now I was alone. The air was cold. There was a full moon in the sky, but it was hazed over by fog.

I untied the cloth with which my wound was bound and plucking leaves and grass, I lightly wiped the wound with them. I then tied it again with the blood-stained cloth somewhat loosely. This enabled me to move about more easily despite the pain.

There area was very quiet. I had no prior knowledge of the road from Nakfa to Afabet. I continued my way eastward following a sandy, dry riverbed that wound between steep, rocky walls so straight they seemed artificial. I heard gunshots about 0300 hours. It did not sound like a firefight, but more like warning shots. I guessed they might be from our troops.

I sped up the pace in as much as I could to try and catch up with the men. Soon enough, I saw a few men sitting by the road.

"How many men are there? Are you the only ones?" I asked them.

"A few men took another direction claiming that they were heading towards Keren. A few others took this road and went ahead of us," they told me, pointing.

Members of the 15th Battalion knew well the roads leading to Afabet and Keren. A few soldiers from my unit had followed them. I told the men sitting there that they had enough rest, that we needed to move on. The enemy was still behind, and we needed to get away from this area as quickly as possible.

"Follow me," I ordered. We moved on to catch up with the troops who were ahead of us. After about an hour, we caught up with the advanced troops. I sat down and took rest till those who remained behind caught up with us. When everybody had regrouped once again, I told them that our best chance was to get off the road and travel across the desert where the enemy was less likely to follow. If we wanted to rejoin our forces alive, this was our best option, but that some of us will not make it.

I told them that on the path we would travel, there is very little water and the desert is very hot. You should not drink any water but only rinse your mouth with water, wet your throat, and spit the water back into your canteen so it won't be empty. Only drink if you can replace the water. When we drink the last drop, we should think of that as a death sentence.

During our travel, we would keep silent. We would see to it that our canteens and our bayonets do not knock together or against our weapons. We would not talk to one another. We would not race, but pace ourselves, moving carefully not to trip or fall to the ground. Silence was imperative. We would use nothing to give off light. After an hour of travel, we would stop and take rest for five minutes. When the rest was over we would rise quickly and quietly, and travel on.

If anyone is unable to keep pace with his team, follow behind slowly as best you can. If we get ambushed, the first team should dash forward and engage them in a hand-to-hand combat to end the ambush. The other teams will follow behind as they have been doing. We cannot defend any ground, there is no point in that. Engage them in hand-to-hand combat immediately, mix up with the enemy to kill them and then move on. We will not bury those who die along the way. We will not carry our wounded. We must keep going until we get back to our lines. Some of us must survive and tell the tale of Nakfa. What we have done here must not die with us.

If ambushes occur repeatedly, the first team to engage the enemy will circle around again and fall in behind the second team who is now the vanguard. In this manner, in repeated waves, we fight through any ambush to the last man. We shall not quit.

We will start moving at 1900 hours in the evening and stop at 0500 in the morning. At 0500 hours, we shall select an area of good concealment and spend the whole day hiding. The selected hiding place will be far from any village, from the main road, from farming areas, from water wells, or from any place frequented by the public. We will post lookouts around our concealed area and once we enter the area there will be no movement whatsoever. We will remain silent and still and rest.

We will defend ourselves if attacked and if we are exposed by some unavoidable situation such as a passerby, a shepherd, or a peasant, we shall keep him with us without doing any harm to him until we leave the place. He will be released as soon as we restart our journey.

With the instructions clear and understood by the men, we began our journey.

On the first day, we travelled until 0500 hours without meeting the enemy. I selected an area suitable for hiding. All day long we stayed there quietly, everybody respecting discipline, making no movement. We saw a lot of people moving about nearby, but none of them saw us. At 1900 hours, when it began to get dark, we went out of our hiding place, and keeping to our plan, we started the second day's journey.

We continued our journey for two hours without rest. At 2100 hours, we took a ten-minute break. After we travelled a short distance, there was a shout from a few men at the back, asking to be given more rest. I told them we needed to follow our plan if we wanted to have a chance of making it. Still they grumbled. Although it was a risk, I allowed five additional minutes of rest. When they were told to resume the journey after the five minutes rest they continued to resist, complaining that they hadn't had enough rest. "What's all the hurry anyway," they asked. "There is no enemy."

"We cannot sit and rest any more. If you don't want to keep moving, you may stay behind." I told them.

We continued forward. They began to cause problems by firing shots and calling out, so after going just a short distance, I stopped and waited for them to come to us.

"Why are you behaving like outlaws? That can only harm all of us and not benefit anyone. From now on, no one shall fire a bullet."

We walked together a short distance. But then they asked again to be given more rest. We ignored their request and we continued our march. Once again, they begun making noise by firing shots and calling out aloud. It was not possible to rest every kilometer and still make our destination. I told them that they may rest if so they wished, but that we are going on. And so we did. We walked until midnight without encountering any confrontation from the enemy. Those marching with me, believing that they were now safe from the enemy, began to sing martial songs and talk loudly. Nobody would listen when told to keep quiet. Some were singing war songs and others mournful tunes. They kept firing their bullets towards the sky, bullets that were so dear to us. We were in a state of anarchy. The troops were out of control. They began to quarrel among themselves over giving or not giving water.

We marched on like this until 0100 hours. Then suddenly we heard an explosion and had hand grenades landed around us again and again. Rocket launchers and machine guns were firing on us. It was an ambush.

I was in the first row in the first team. When the team heard the sudden sound of shots and was about to turn and retreat I shouted, "Go forward."

When they saw me firing and flinging myself into the enemy ambush, they followed behind me. It was a hard fight. Eight of us, in the first row, were engaged hand-to-hand. Three of our eight men were hit and fell to the ground. The remaining five of us broke through. My rifle had been struck by a bullet and would now fire only one bullet at a time. The other teams stepped over the dead bodies of those who fell and followed behind us. The last team, believing that we had been captured, retreated.

We captured four light weapons and one machine gun in the ambush. We regrouped and continued our journey. We faced another ambush in less than an hour. Again, I lined up with the first team and we broke out. All in all, we had five skirmishes with the enemy before daybreak. Many heroes fell in these fights. Those of us still alive stepped over the dead bodies of our fallen brothers, climbed the slope of Kukub, and reached the plains of Afabet at dawn. We were still 15 kilometers from the town of Afabet, where the camp of our forces was located.

There was no water in our canteens. There wasn't any kind of water source in the surrounding areas. Our food was also depleted. We'd reached a stage where we simply could not withstand the thirst and hunger. Therefore, we decided to run the risk of traveling by day to arrive sooner, rather than spend the day in hiding while desperately thirsty and hungry. Some of the men could hardly walk. Others could not even talk because of severe dehydration. There were many who were left behind. I was concerned that they might die of dehydration and fall to hyenas and vultures, or fall into the hands of the enemy.

After about four hours of traveling, we could see Afabet about 3 kms away. We felt as though we had joined our forces already. After a short distance, we reached an area with trees. The troops put down their weapons and lay down in the shade, totally stretched out. I chose two men who were not as exhausted and asked them to walk to the 16th Battalion to ask the commander to send us a vehicle with some water and food. Soon after the two men left, some gunshots were fired at us from a nearby cemetery. Our troops who could hardly breathe, talk, and move were suddenly revived. They raised their arms and began to fire back, making use of cover and concealment tactic. Some of us assaulted the enemy from the flank while others gave fire cover, forcing the enemy to retreat towards Kukub. The troops of the 16th Battalion thought that the gunshots they heard was a fight among the separatists themselves.

The difficult passage between Kukub and Afabet where we broke through the ambush of the separatists. This picture was taken 10 years later, in 1987, in the place named by both patriots and the enemy, as "Tesfaye's passage."

CHAPTER 16
Rejoining Our Forces

THE MEN OF the 16th Battalion and their commander were shocked when they found that the two men I sent to get us a vehicle with food and water were from Nakfa. It was unimaginable to them that anyone could have survived that six-month siege. Putting aside his disbelief, the commander sent us a vehicle loaded with food and water. The provisions revived us considerably, and we joined our forces in Afabet.

We later learned that the group that went to Keren broke all the principles of escape and evasion, moving during daylight, eating salty foods, stealing from civilians, and letting their guard down. They took goats from peasant herders, then slaughtered and ate them. They also approached villagers for food and water and ate what the villagers provided them.

While some of the villagers treated them kindly and made them tea and fed them, other villagers slipped into the woods, uncovered the weapons they had buried and waited to ambush the soldiers.

When some of our men left, they were ambushed, and were massacred with mines, grenades, and machine guns. The separatists went around shooting into the fallen bodies to make sure that they were dead. Some local villagers collected their weapons and equipment, stripped them of their uniforms, and left the men naked and unburied. Leaving the dead soldiers to wild animals and vultures, they returned to their villages singing and boasting.

This was learned from an Airborne soldier, named Tadesse Ferisa, who'd played dead and observed what was going on. Twice he luckily escaped death as the separatists went back and forth shooting into the fallen men.

When it started to get dark, he pulled himself out from beneath a pile of dead bodies and resumed his travel. Learning from their earlier mistakes, he travelled only at night and hid himself in the bushes during the day. Tadesse Ferisa joined our forces in Keren after five days, living on figs and leaves, and using airplanes flying to Keren for his navigation.

He reported that he had started with a team of twenty and that some members of the Airborne unit had proposed that they should travel using the escape and evasion tactics. However, members of the 15th Battalion argued, saying they knew the surrounding area, that the inhabitants of the area were peaceful people, that they would face no problems. If they travelled by day, they could rejoin the army in Keren sooner.

"Members of the 15th Battalion outnumbered us. We gave in to their arguments and travelled by day. We moved into the villages in search of food and water and fell into the hands of their fifth columnists," he relayed regretfully.

I sent a telegram to the 2nd Division Commander, informing him that we had reached Afabet, and requesting that a ground contingent be sent back to assist our men who had been left behind. The response telegram said that helicopters would be dispatched to fetch the men who'd been left behind. As for the soldiers who had survived the siege of Nakfa, they should be ready because helicopters would be sent to bring them to Asmara. Major Tesfaye Habtemariam should come in the first helicopter.

When we broke out of the Nakfa siege, it was not possible to know exactly how many men died during the breakthrough, how many were wounded, how many were captured by the enemy, or how many surrendered to the enemy. Only 72 of us reached Afabet; 23 from the Airborne unit and 49 from the 15th Battalion.

After a few hours, two helicopters arrived at Afabet from Asmara. I was told that a Dakota aircraft would transport the troops from Afabet to Keren and a transport plane would take them to Asmara.

The 2nd Division Commander ordered me to fly to Keren in the first round of helicopter flights, then head from Keren to Asmara. I refused to leave Afabet before making sure that all my troops left first. I received another message, which stated that I not be left behind in the second round. Again, I refused until the last of the remaining troops had been airlifted to Keren. I left in the last round of flights to Keren, then flew to Asmara.

When I arrived at Asmara airport, I was met by several military generals and high-ranking officers from all the branches of the military, as well as the chief administrator of the region and other civilian authorities. They were waiting in front of the terminal. I thought they were there to welcome some high official from Addis. I descended from the plane and walked in the opposite direction from where they were standing.

I turned back when I heard a familiar voice shouting, "Major Tesfaye! Come over here!"

It was Col. Fikru Wolde Tensae's, Chief Administrator of Eritrea and Commander of the Northern Front, whom I knew very well from my cadet years in Harar. I walked towards him and gave him the rifle salute. He choked with tears as he greeted me. He said he was extremely happy and grateful that I was still alive. All the high-ranking military officers and civilian officers, some whom I knew personally and some I did not, greeted me, one after the other. Then I realized all the officials had gathered here to welcome me.

When the welcoming ceremony at the airport came to an end, I was taken to the Kagnew Military Base escorted by a long procession of vehicles and motorcycles. A large villa was reserved for me on the base. I was given new clothes, new shoes, underwear, socks, and towels.

After I took a warm bath and got dressed in the new clothes, Col. Yilma Gizaw, Operations Officer of the Division, came over to my villa to take me to the officers' mess hall. Col. Yilma Gizaw had been my teacher during my cadet years and he was one of the first batch of graduates from the academy. We knew each other very well.

Special preparations had been made in the officers' mess hall and I was given a dinner reception in the presence of officers from all the armed forces and the Division Commander himself. After the dinner reception, I went back to my villa to rest. I could not sleep as I was still visualizing in my mind the struggle of the last six months. Feeling the pain from the bullet still inside my right leg, brought back the memories of Nakfa and I had trouble falling to sleep. Finally, night was replaced by dawn. In the morning, I went to the hospital to have the bullet in my leg removed.

Later that day, the Division Commander ordered that all of us, officers and NCOs who broke out of the Nakfa siege, should receive a thorough, medical checkup. I was happy to see my soldiers at the hospital. I found out that I weighed only 48 kilograms, which meant I'd lost 8 kilograms since our jump into Nakfa. After completing a series of medical examinations which took a couple of days, we flew to Addis to see our families

The day after I arrived in Addis, I delivered an hour-long report on Nakfa to the President of Ethiopia, and members of the Derg in the presence of the Chief of Staff, the Commander of Army, and officers from many different departments. In the report I submitted, I focused on the fate of our forces in Afabet and Keren.

I pointed out that the breakout of the 15th Battalion and Airborne unit from the Nakfa would create an opportunity for the separatists, and that Afabet and Keren would be their next targets. I proposed that Afabet and Keren should be reinforced immediately with men, weapons, and begin securing defensive positions on the surrounding high ground. If reinforcement was not possible, I recommended a strategic retreat from Afabet, and moving those forces to reinforce Keren and make a stand there. This would prevent the loss of our men and weapons in both places.

I sensed a unanimous acceptance of my proposal and felt that the authorities would take immediate action. After I was done with my report, I was told I could go on a one-month leave to spend time with my family in Debrezeit.

Less than 15 days later, on April 7th of 1977, the separatists wiped out the force in Afabet soon after the first reinforcement troops from Asmara were flown to Afabet. The pilot, carrying the second load of reinforcement troops from Asmara to Afabet, could only see black smoke as he looked down at the military camp of the 16th battalion from the air, and there was no response to a radio communication to Afabet. The pilot had to return to Asmara with the reinforcement troops and the weapons he carried.

Then once again, timely action wasn't taken to reinforce Keren and prevent another disaster like the one that just occurred in Afabet. The separatists, using the men and weapons from the troops they captured in Afabet and Keren, began heading towards Asmara. In a short time, the separatists had constructed a strong trench that could withstand shells from heavy guns and aerial bombardment just 10 to 15 kilometers from Asmara. To build the trench, they used the materials taken from the railroad that ran from Asmara to Akordat.

During this time, my family leave ended, and all of us who broke out of the Nakfa siege received promotions. I was promoted to the rank of lieutenant Colonel, only four months after I had made Major. Typically, this takes five years to achieve. I was also appointed Commander of Air Security. A few days after I started serving in this capacity, I was called to the Derg office.

I was told that I had been appointed commander of the Special Force battalion and was ordered to go to Debreberhan immediately and take over the command. I arrived at Debreberhan the next day, which was the beginning of August 1977. I took over command of the battalion and soon thereafter, was told to report to the office of Derg in Addis. When I arrived, I met with the First Vice Chairman of the Derg, Mengistu Haile Mariam. We talked about the seriousness of the current situation and what to do.

It has become necessary to establish three Para-Commando Brigades.

While still maintaining your post of Commander of the Special Forces, you shall, together with Major Shibabaw, and Major Shiferaw Worku, work on recruiting academically qualified students from the militia at Camp Tatek. When you finish recruiting your men, we will identify a good location for the training center. When the recruitment is complete, the recruited men will be grouped in three brigades, and each named, 1st, 2nd, and 3rd Para-Commando Brigades. I would be appointed commander of the 1st, Major Shibabaw Zeleke of the 2nd, and Major Shiferaw of the 3rd Para-Commando Brigades.

I left for Awassa Lake with the 1st Para-Commando Brigade to complete the first level infantry training. The 2nd Para-Commando Brigade, under the command of Major Shibabaw Zeleke left for Awash-Arba. The 3rd Para-Commando Brigade, under the command of Major Shiferaw Worku, left for Legedadi.

As of September 1977, the situation of the country had become precarious. The peasants set aside their ploughs and yokes and took up their guns to fight the separatists. I guessed that no one would question whether the Para-Commando Brigades, though it was to be part of the regular army and a future strategic force, would stand in reserve or extend its training given the current situation.

While following the regular training program prepared by our superiors, I came to realize that I needed to give my men more pressing basic training such as how to aim correctly, shoot at one's enemy, and kill him. We started this training in October of 1977.

At 1900 hours, during our training on November 23rd, a Derg official came to our training center followed by scores and scores of vehicles from Addis. He ordered us to stop training and for the troops to get into the vehicles and proceed to Addis immediately. I was not surprised by this. I informed the Derg official that it would require some time to gather all the men and issue them their weapons, and that the weapons needed to be uncased and cleaned of the grease they are packed in.

The Derg official replied, "This is a mission to save a human life!"

While I did not know what the situation was, I did know that it was foolish to send troops into battle without properly arming them. I assured him that the troops would shortly be ready for the mission.

CHAPTER 17
The War in the Eastern Front

As ordered, all four of my battalions ceased training and returned to the camp immediately. The Derg official arrived at our camp with hundreds of transport vehicles to take the troops to Addis. The soldiers would fly from Addis to Dire Dawa. They then would travel by road to join the ground forces fighting in eastern Ethiopia.

Our armory in Shashemene, where the troops would draw their weapons, was some 24 kilometers from Awassa and could only handle one battalion at a time. We lost no time in getting the first battalion on the vehicles. The troops were supplied with a quickly prepared meal. While three battalions ate their supper at Awassa Lake, the first battalion began drawing their weapons from the armory. Each gun had to be uncrated, cleaned with kerosene, and dried. The magazines had to be loaded with bullets. It was midnight by the time the first battalion was armed and ready to start the journey from Shashemene to Addis. The other battalions followed next, with the fourth battalion finished and leaving for Addis at 0200 hours. I arrived at Bole International Airport in Addis at 0600 hours with the fourth battalion. It was November 24th, 1978.

We boarded a Boeing 707 Ethiopian airplane headed to Dire Dawa. The trip took less than one hour. We completed our air travel to Dire Dawa and immediately continued our journey to Harar by road. At the crossing of the way from Addis to Harar, and the road from Dire Dawa to Harar, we came under heavy fire from mortars and large-caliber machine guns. The enemy was trying to delay our forces. The drivers of our trucks were civilians, and I was worried that they would be frightened and unwilling to continue when they heard the shots. I ordered the drivers to keep driving as long as they could. We did not dismount from the vehicles to return fire on the enemy. Our mission was not to engage with the enemy on the way but to save Harar. Some soldiers were wounded in the vehicles, but we continued our journey to Harar regardless.

The shots from the enemy intensified as we entered Harar. We stopped at Aboker and dismounted from the vehicles, took cover and began to return fire. I had my men set up a defensive perimeter and made my way to the command post for orders. Mortar rounds continued to fall on our position. On my way to the command post, I met the operations officer for the eastern command, Colonel Mesfin Gebrekal. He was pleased when I told him that I was there with my troops. He ordered us immediately into battle at Hakim Gara.

"Colonel Tesfaye," he said. "The enemy is about to take the town of Harar. I have no specific orders to give to you. Starting here at Aboker, try your best to take Harar town back from the enemy. That is all for now." With that, he left.

I immediately ordered one battalion to advance to the north and seize control of Aboker Hill from the enemy. I ordered the second Battalion to proceed from the Teachers Training Institute towards Kelad Amba. As the area they would be pushing through could not accommodate more than two battalions at the same time, I ordered my other two battalions to follow behind as reserves. The Brigade Operations Officer would coordinate the battle by advancing with the second Battalion. I moved forward with the battalion pushing toward Aboker Hill.

Aboker Hill overlooked the town of Harar. The enemy had control of the hill, near St. George Cathedral, and was lobbing mortars into the town. The mortars were killing civilians, and I knew I had to neutralize that situation immediately. Taking Aboker Hill meant freeing Harar.

After two hours of fighting, we took Aboker Hill. Taking the hill exposed the enemy to greater danger because the land to the north of Aboker hill through which they had to move was flat. The fighting was intense, and many lives were lost. Finally, I gave the order to flank St. George Cathedral. Using suppressing fire, we launched a surprise attack from both the west and east. The enemy, surrounded by our troops, abandoned the hill and retreated towards Kombolcha.

The enemy units holding Kelad Amba, seeing that Aboker Hill was lost to them, feared getting cut off and fled. I ordered one battalion from the reserve to join the battle and chase the enemy retreating towards Kombolcha. With support from our Air Force, we pushed the enemy to Kombolcha and killed many in the process.

At 2000 hours, we had Kombolcha under our control and liberated our forces who'd been surrounded there. We set up our defenses at Kombolcha and spent the night. When I reported to the operations officer of the Eastern Command that Kombolcha was now under our control, he would not believe me. He said that the most he had expected from us was to free Kelad Amba. On December 4th, 1977, Colonel Mesfin, Operations Officer of the Eastern Command, arrived from Harar just as we completed our preparations to proceed toward Ejersa.

Colonel Mesfin informed us that the Commander of the Eastern Front and the Minister of Defense, as well as members of the Derg, were extremely pleased by the success of the 1st Para-Commando Brigade in liberating Harar. He returned to his command post in Harar after giving the following instructions:

"The 2nd Para-Commando Brigade that advanced from Hakim Gara, south of Harar, should stay where it is and hold its position due to increased pressure by the enemy. The existing troops, together with the 1st Para-Commando Brigade, will strengthen the defense of Kombolcha. Lieutenant Colonel Tesfaye HabteMariam will maintain command of the 1st Para-Commando Brigade and coordinate the security of the Kombolcha zone."

Kombolcha is on a higher ground than the surrounding area and appears to be a suitable defensive position, but it sits in the shadow of Ejersa Goro mountain, located about ten kilometers to the east where the enemy positioned artillery with a range of 25 to 30 kilometers. Therefore, the enemy could attempt to dislodge the patriotic forces using heavy weaponry rather than deploying its infantry. Consequently, we needed to construct strong trenches.

Most of the troops had no experience and had not completed basic training, so they did not know how to construct trenches. With the help of combat engineers and my close supervision, having a lot of experience from my time in Nakfa, the troops worked day and night to build strong defensive positions. The enemy poured fire on us from artillery at Ejersa Goro. The shelling compelled us to work only at night. It was not possible to move about during the day to collect the materials we needed. Several people were injured by artillery fire during this project. Once we completed the trenches, we were safer.

The militia refused to construct trenches. Consequently, as soon as they started receiving artillery fire, they fled from their positions towards Alemaya where they hid in sorghum fields. Their retreat opened a big gap. Some of the troops went AWOL. That evening some returned to Kombolcha to retrieve their rations. When they were denied their rations, they demanded food at gunpoint.

I gathered the men and explained to them that leaving one's defense position during a battle and threatening their comrades at gunpoint was not allowed. This act was punishable by death. They jeered, saying they had come to fight enemies with rifles and not to fight against God's wrath, the thunder of artillery. The 1st Para-Commando Brigade had to dig all the trenches. When the trenches were completed, they still refused to stay in them during the day. I told them no soldiers in the trenches had been injured, and that they would be safe in these positions too. Still, they refused.

From this point on, those who left the trenches would be considered an enemy in our eyes. We lined our positions with explosives. Anyone who fled would be responsible for his death. We lined the exit roads with boobytraps.

"You must stay in the trenches," I ordered.

As a preliminary measure, engineers were instructed to prepare lower-level explosives and to set them off when they saw militiamen trying to escape. Still, the militia attempted to flee, and when they did, they set off a series of explosions. The commotion startled the militiamen, and they returned to the trenches, afraid for their lives. Eventually, they adapted to life in the trenches.

Sorghum fields covered most of the land around Kombolcha. The peasants put up watchtowers in their fields to chase away birds with slingshots. When the birds let them relax, they descended from the towers and pulled weeds. They cut the worm-eaten, seedless sorghums, so they didn't poison the healthy plants. We often saw peasants passing before us in the evenings on their way home. They would come and go from their fields right by us without a problem.

Enemy scouts, some with the help of peasants, and others disguised as peasants would come right up to our positions to select and mark strike targets during the day, then return to their side in the evening. These spies waited for me, watching my movements as I supervised and gave directives to the troops. They began taking shots at me.

Initially, I took the fire for coincidence, not suspecting it to be from enemy snipers. To make sure, I left my retinue behind and went about with only a short antenna radio operator. I realized then that they were firing at me. I now suspected that there were forward observers following my movements and watching to kill me.

To verify my suspicion, we stopped the peasants from passing through. Earlier that morning, before daybreak, we'd sent out scouts to hide and observe. Later that morning, as it was getting warm, we saw two men dressed like peasants. One peasant had a sickle in one hand. In the other hand was a radio handset. The radio was under his clothes. I realized that the peasants were sending messages about our camp to the artillery and mortars in the hills.

A heavy gun is much more effective with the help of a forward observer. To zero in on the target, it must get corrections from the forward observer. It was clear that the peasants were serving as forward observers. After we captured the agents, the frequency and accuracy of the shelling lessoned.

A second unsuccessful attack ordered by the Somali government led to a substantial loss of life. Typically, they tried to inflict damage from a distance by using heavy guns, tanks, and anti-tank weapons. Somalia could not carry out a continuous war due to economic limitations. They needed to stop the fight to resupply themselves with supplies such as food, ammunition as well as replace the soldiers who had been either killed or wounded. The Somali government realized that they would face severe administrative and logistic problems if they continued fighting so far from their main base. They developed a plan for a two-pronged attack and began to try their luck one last time. One prong would start from Ejersa Goro and attempt to eliminate the Ethiopian forces entrenched at Kombolcha. The other prong would begin from Fedis and advance to eradicate the Ethiopian troops positioned at Hakim Gara. If these actions were accomplished, they would then drive north to take Harar under their control and cut off the troops stationed at Kore. They prepared their forces, and on January 14, 1978, launched both attacks simultaneously.

They shelled our positions intensely for half an hour. It would seem incredible that anyone could survive such a barrage. However, due to the strength of our trenches, the troops at Kombolcha did not budge. We merely waited, ready for the enemy to approach. When the shelling stopped, the enemy fired on our positions with small arms, tanks, and anti-tank weapons. They pounded our trenches to demolish them. When they thought that they had finished the job, their infantry came close and began throwing hand grenades at us. Then, with fixed bayonets, they charged towards our positions. Our soldiers, who had been quietly waiting, opened fire, cutting them down, one after another. The bodies of their dead were piling up in front of our trenches. Eventually, those who survived turned their backs and retreated. This first assault by the enemy came to an end at about 1000 hours. The enemy troops regrouped and launched their second assault at 1200 hours. My men, encouraged by their success earlier, again waited patiently until the enemy drew near. They waited until the enemy was close, and then opened fire, once more forcing them to retreat.

At 1400 hours, the firing stopped. We heard occasional blasts from the bigger guns. I went out and surveyed the surroundings. I told my commanders not to relax as the battle was not over. The wounded were taken to the hospital for immediate treatment. The burial ceremonies for the dead soldiers would take place only in the evenings. We needed to be alert because the enemy could, at any moment, launch another offensive. While I was still giving my orders, an artillery shell landed roughly five meters away from where I was standing.

The impact of this explosion was tremendous. The heat from the blast was intense. Big trees were splintered into pieces. Some men lost their hearing due to the shock wave. Others lost their sight due to the flash of the explosion. Splinters tore into body parts.

A shell fragment, searing hot, gashed the left side of my head. The force of the explosion threw me up into the air, and I landed head first in a nearby trench.

The trench collapsed from the explosion after I fell in. I was buried alive. I was unconscious for some time and had no awareness of what happened next. Later, survivors told me the following.

"We continued the battle believing that you were dead. We reported your death to the Commander of the Eastern Front, to the H.Q. and the Ministry of Defense Operations Officer. Your death was then reported to the 1st Deputy Chairman of the Derg, Mengistu Hailemariam. Two hours, after the battle was over, we selected your temporary burial place, so your family could easily recover your remains to bury you later. We took you out of the rubble, and as we were taking you to your burial place, one soldier noticed you were alive."

"His body is still warm, and he is still breathing, although faintly," he yelled. We did not believe that you would survive, but we decided to send you to Harar 3rd Division hospital.

When my body arrived at the hospital, I had surgery. The doctors turned my body to examine my condition and seeing that I had lost a lot of blood, concluded that I had a severe brain injury.

"We don't believe that this man has any chance of survival. We cannot waste our time on him. You had better bring in someone who has a chance."

They set me aside. Luckily, one of the doctors insisted that the wound be washed first, and then he re-examined me. When the wound was cleaned, they saw that the shell fragment had taken off a big slice of my skull but had left behind a thin layer that kept my brain intact. The doctors were surprised. They immediately began surgery to remove the fragments that were embedded in my skull.

Later, I was told that the field surgeons did what they could to extract only the large fragments that were life-threatening. The other pieces required specialists with advanced implements. If I survived, I would have to be sent abroad where these were available.

As this happened, my wife, defying the cold, the rain, the heat, the hunger, prayed for me day and night. She went from one church to another, crying endlessly, fasting and begging God that I return alive and without any body parts missing. Perhaps, I was saved by luck. But I believe I was miraculously saved because God heard my wife's prayer and saw her tears, her fasting, and worship.

The one hospital that received the most significant number of the wounded was the 3rd Division Hospital where I was being treated. Some lucky patients could move and go out to the yard to get sunlight, while others could not even sit up. Many needed crutches while still others were confined to wheelchairs or were carried out on a stretcher. There were so many wounded men in the hospital that it seemed to me that no one could be left at the frontlines to fight. The horror of war was evident as I walked through this hospital. It was horrifying for me to see the disfiguration and dismemberment that the weapons of war had caused. It was during my stay in this hospital that I came to realize, more than ever, that war leaves an invisible scar that can pass from generation to generation, a scar that cannot be quickly erased. The greatest measure of the sacrifices made by the Ethiopian people, by the regular army and the militia, to protect the sovereignty of their country, could be found in the hospital in Harar.

Although I made every effort to keep my condition from my wife, she found out. She arrived at the hospital one week after I was admitted.

She flew from Addis to Dire Dawa and drove by car from Dire Dawa, within range of enemy fire, all the way to Harar. The intense battle raged on, and heavy guns could still be heard when she arrived.

By the time, my wife got to the hospital; my condition was improving due to all the incredible medical attention I received. I could stand and moved about reasonably well by myself without support from anyone, although I admit I was still a bit unstable. I talked, although my speech was incoherent. I regained my memory gradually every passing day.

The right side of my body remained numb and insensitive to cold or heat. I could not hold objects with my right hand. Any objects I tried to hold fell to the ground without my awareness. I could not turn my neck sideways. If I wanted to see something at my side or behind me, I had to turn my entire body in that direction. My neck could not yet bear the weight of my head, so I was given a brace to put around my neck, and this rested on my shoulder to support my head.

I was spared God's final sentence, the termination of my earthly contract. Still, the doctors were less than optimistic that I would ever fully recover, although they assured me that I had escaped death.

My wife stayed by my side for one week as the battle raged uncomfortably close to her. She suffered a lot because of me. She finally returned home after the doctors reassured her that I would rejoin her at home in just a few days, once I was strong enough to travel.

The Ethiopian forces planned to launch a counter-offensive at the beginning of February 1978. The enemy somehow got word of this and, for the third time, began an assault on the Kore, Kombolcha, and Fedis fronts. For the 3rd time, the brave resistance of our troops repulsed their attack. The loss of life and property was significant.

The relationship of the Somali government with the Soviet Union was severed, and all the support they were getting in the form of weapons and ammunition came to a stop. Although they still had weapons, they had no ammunition or spare parts. Western nations, especially the United States, promised Somalia that they would supply them with all the required arms if Somalia discontinued its relationship with the Soviet Union. But, the United States did not keep its promise even after the Somalians threw out Soviet personnel and terminated relations with the USSR.

Somalia later managed to procure weapons and ammunition by purchase or aid from various countries, such as Egypt, Iraq, Belgium, Saudi Arabia, Spain, Italy, and Pakistan. But the weapons and ammunition, as well as spare parts, procured through purchase or aid from different countries, were unreliable. In the meantime, Somali troops received orders from their superiors to shift from the offensive to a defensive posture.

The Somalis, to defend the Kore front, which they believed would be the main line of attack by Ethiopian troops, established their line of defense at a place on the west side of Jijiga town. Jijiga is located on the east side of the Karamara mountain, straddling the Harar-Jijiga road that cuts through the Karamara mountain.

The enemy, wishing to hold on to Jijiga town for good, constructed a strong trench along its defensive perimeter. In front of their positions and in all the places they thought we might advance, they buried anti-tank and anti-personnel mines, prepared barriers, and waited.

After stopping the third assault by the enemy, the plan was to attack before they got a chance to regroup and reestablish themselves once again. Our ground forces went directly into attack mode to chase and strike the enemy. It was decided that the assault would be a four-pronged attack, hitting Dire Dawa, Kombolcha, Kore, and Fedis.

The troops in holding our positions at Kore would remain in where they were but would launch a diversionary attack from there. For the soldiers starting from Dire Dawa, the land they were using was suitable only for the mechanized forces. They would be reinforced with tanks, armored cars, and military vehicles. When they finally started from Dire Dawa, they eliminated the enemy found in Galdessa, then proceeded towards Arabi, Gugti, Teferi-Ber and then to Togochale. They regrouped at Togochale, and then stopped. They destroyed the enemy coming from Hargeisa to reinforce Jijiga, as well as those fleeing from Jijiga and trying to enter Hargeisa.

The troops attacking from Kombolcha were reinforced with infantry and provided air support and artillery cover. They would cross the Fenala river and eliminate the enemy forces found at Ejersa Goro and Fugna Bira, and take this area under control. Tanks and armored cars would also reinforce the troops at Fugna-Bira. They would then assault Jijiga from the north and take control of this region.

The troops advancing from Fedis, since the terrain leading up to the Fafem River is not suitable for mechanized troops movement, would be assigned an infantry unit and supported by air and artillery, and would eliminate the enemy force found at Fedis. As soon as they crossed the Nuram river, using tanks and armored cars, they would proceed toward the Jijiga - Degehabur road and place it under their control.

All units, to prepare for the successful completion of this battle plan, after receiving their orders, moved in secret during the night to their starting points. Orders were given so that the battle would be executed in the following sequence:

> *First, the troops at Kore front must firmly hold on to their defensive position and simulate an intense assault, thereby compelling the enemy to put all its strength on the Jijiga front. To prevent the enemy from strengthening itself by transferring troops from Jijiga and sending them to reinforce other fronts, harass them by attacks from the air and artillery. When the soldiers at Kore begin their feint, simulating an attack to draw the attention of the enemy towards Karamara, the mechanized at Dire Dawa shall move quickly and eliminate Galdessa, Arabi, Gugti, and Teferi-Ber, and put them under their control. They will then move swiftly to Togochale and from there shall cut off the enemy at Jijiga to stop reinforcements from Hargeisa reaching Jijiga. Once they are encircled, the enemy troops will be asked to give themselves up peacefully. If they refuse to put down their arms, they will be forced into submission.*
>
> *The troops lined up at Hakim Gara, in similar hours as those at Dire Dawa, shall launch an assault at Fedis. As soon as they put Fedis under their control, they shall advance, crossing the Nuram river, and divide their force into two groups. One will flank Jijiga and head north. The second team will bypass Jijiga to its left and control the Jijiga-Degehabur road. It will stop the enemy's reinforcements coming from Degehabur. The enemy troops, fleeing*

from Jijiga and attempting to enter Mogadishu via Degehabur, will be asked to put down their arms and give themselves up in a peaceful manner. If they don't, they will also be forcibly taken under control.

The troops lined up at Kombolcha will attack and take over Ejersa Goro and Fugnan Bira. They will then turn towards Jijiga, launch an attack from the flank in the direction of north to south, and together with our forces assault the enemy in the southern area, and shall put Jijiga under their control jointly.

Following this directive, and in keeping with the sequence described, the Ethiopian forces will begin their combined attacks at the appointed time.

The troops who started from Dire Dawa entered Teferi-Ber via Arabi. Next, they entered Togochale and successfully cut off the enemy. The force from Kombolcha took Ejersa and Fugnha Bira under control and then turned south towards Jijiga. The troops who started from Hakim Gara put Fedis under their control, crossed the Nuram river, and placed the Degehabur road under their control. The enemy troops, realizing that they were encircled and had no way of escape, put down their arms and surrendered.

The Somali government, realizing that its troops were trapped, appealed to both the United States and the USSR for help. Both governments urged an immediate ceasefire. They demanded that the Ethiopian forces stop their pursuit of the Somalis and hold their present position and that the Somali army immediately withdraw from Ethiopian territory, moving out within 24 hours.

The joint proposal was acted on right away. The Derg accepted the ceasefire because it had other military emergencies to deal with, especially in Eritrea. Putting an end to the war in the Eastern Front meant that the Derg could turn its full attention to the north.

The ceasefire happened at a time when our military was on the verge of complete victory. We had to let our enemies go, waving white flags of peace as they left Ethiopian territory.

The Ethiopian forces opened the Jijiga to Hargeisa motor road. The enemy troops fled Jijiga to Hargeisa on the very day of the ceasefire. Along the way, they dismantled roofs, doors, and windows, and loaded these items onto their vehicles. Ethiopian soldiers were greatly disappointed that they were not allowed to avenge the blood of their brothers and honor their Ethiopian people. Some even shed tears of remorse.

Once the enemy forces returned to Hargeisa, the Ethiopian troops entered Jijiga town and raised the Ethiopian flag on Karamara Mountain. All the Ogaden cities, towns, and villages; Degehabur, Qebridehar, Aware, Wordaree, Gode, Mustahil and Geladin, were now liberated from the enemy, and the Ethiopian flag waved over all of them.

The Ethio-Somalia war, which took place from July 1977 to February 1978 had finally come to an end.

In the wake of the war in the east, the Ethiopian forces, with the slogan, "The victory in the East shall be repeated in the North," left sufficient troops to ensure security in the region and turned their attention to the North.

For my part, following the decision of the medical doctors, I was released from Harar hospital and spent two months recovering in the care of my family at my home in Debre Zeyit. After staying in Debre Zeyit for two months, I was sent along with other military officers, to the Moscow Frunzi Academy to attend a training course for brigade commanders and staff officers along with medical treatment.

The Outcome of the War

It is difficult to account for the impact of the war between Ethiopia and Somalia, regarding the loss of human life and destruction of property.

Both sides reported staggering numbers of deaths and damage done to the enemy while minimizing the number of deaths and destruction to their forces. The figures of casualties listed below are taken from an article written by a British journalist named, P.S. Gilkes, entitled, "*The Ethiopian troops in Ogaden and Eritrea, 1974-1984*." P.S. Gilkes noted that his figures were not 100% accurate.

The casualty figures in both countries, as issued by the Somali government

	Killed	Wounded	Captured
Casualties on Ethiopian Side	23,000-24,000	-	-
Casualties on Somalia's side contains three different versions due to its inconsistency.	1st Version: 7,000 2nd Version: 11,000 3rd Version: 20,000	15,000	-

The casualty figures in both countries, as issued by the Ethiopian government

	Killed	Wounded	Captured
Casualties on Somalia's side	18,000	-	1785
Casualties on Ethiopian Side	700	1,500	-

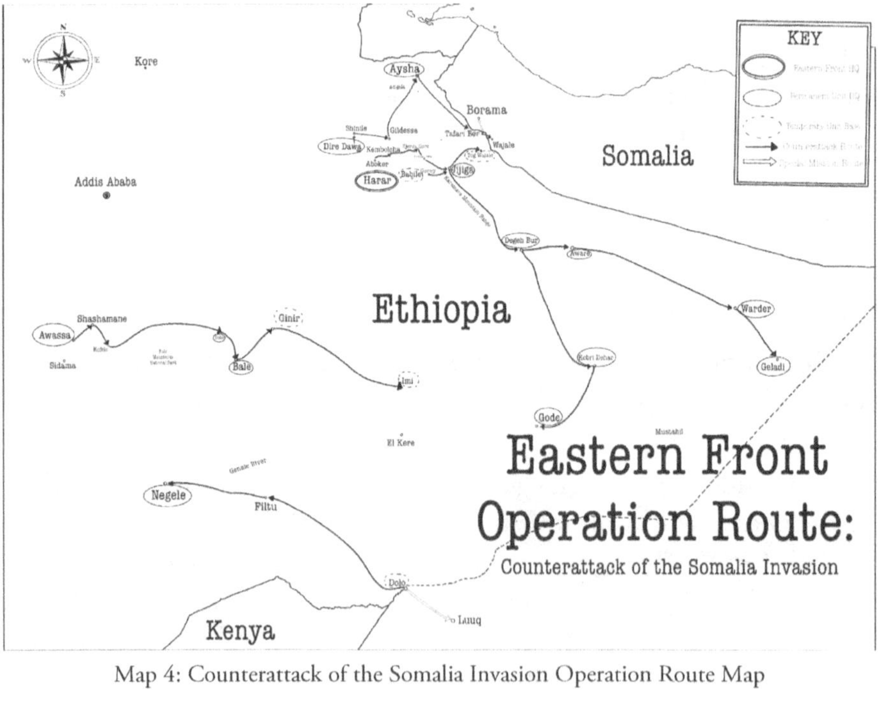

Map 4: Counterattack of the Somalia Invasion Operation Route Map

CHAPTER 18
The War in Eritrea

THE UNITED NATIONS decided that, after being colonized by Italy for fifty years, Eritrea should fall under British military administration. Great Britain planned to cut away the western Eritrean district and make it part of Sudan, which was a British colony at the time.

There was a group in Eritrea called Mahibere Ager Fikir. This group had resisted the Italian occupation of Eritrea and was opposed to Eritrea being split off and made part of Sudan and controlled by the British. Mahibere Ager Fikir wanted Eritrea returned to Ethiopia, its motherland.

Mahibere Ager Fikir was a real thorn in the British side, because of its objection to being annexed by Sudan and British control. Other political parties that opposed Mahibere Ager Fikir and supported the British expansion plan were established with the help of Britain.

These were The Progressive Party, which supported the Eritrean status as a British protectorate until Eritrea become independent and the Muslim League, which supported the idea that the Eritrea western desert should break away and be annexed to Sudan because the inhabitants shared the same language, religion, and culture with the Sudanese people.

The Liberal Party wanted Eritrea to be returned to Italy as a colony. However, before the British plan to annex Eritrea into Sudan could happen, the United Nations decided Eritrea would become a federated part of Ethiopia as of September 11, 1952. Great Britain's plan to control Eritrea had come to an end. Accordingly, Tedla Bairu, leader of the Mahibere Ager Fikir party, was appointed President of Eritrea, under the federal government, and the leader of the Muslim League party, Sheikh Ali Mohammed Mussa Radain, was named Vice President.

Ethiopia and Eritrea were administered under this federal arrangement for several years. On January 9th, 1958, Eritrea became the 13th region of Ethiopia, by the decision made by the national parliament. The Eritrean President and Vice President, as well as their collaborators, believed that under the united government, the Eritreans would have administrative control of the region. When they discovered that the man appointed as chief administrator of the region by the Ethiopian government was Ras Andargachew, who was not from the area, they were disgruntled and exiled to Egypt. They formed a political party known as the Eritrean Liberation Front (or Jebha) and welcomed all Eritreans who wished to join. Jebha requested aid from many Arab countries. After a short period, Jebha established its military training school in Khartoum, Sudan.

Jebha put out a call to all Eritreans of proper age and good health, asking them to come to Khartoum and get military training and to be ready for an armed struggle. Those who were not of age, or not fit to take part in the armed conflict, were asked to provide organizational support, supply intelligence, and aid the wounded.

In response to this call, Eritrean men, women, boys, and girls left their schools, jobs, or businesses, and made their way to Sudan to join the armed struggle.

Soon after Jebha's recruits completed their military training in Sudan, they infiltrated the western desert of Eritrea and began attacking recreation centers, marketplaces, schools, and hospitals. They killed and maimed innocent citizens with no political involvement of any kind. They placed anti-vehicle and anti-personnel mines on motor roads and inflicted damage on innocent civilians and their property. They stopped buses and robbed the passengers, taking money and jewelry. The Jebha quickly disappear from the scene any time Ethiopian forces approached. They would lay low until the Ethiopian military moved on. For many years, Jebha continued to carry out this type of hit-and-run attacks.

Over time, the Jebha fighters became more experienced and obtained better weapons. They shifted to more sophisticated terror attacks. They began to use explosives to destroy bridges and factory generators. Where, in the past, they would run away at the approach of Ethiopian forces, they now began to stand and fight.

Jebha's use of explosives to attack our infrastructure increased daily. The increasing attacks required that troops be assigned to most of our infrastructure to protect them.

The Ethiopian military used many of its units to protect our infrastructure, essential business organizations, and well-known personalities. This shift created a shortage of offensive troops and allowed Jebha to move around without fear.

The places that needed protection were many and often were scattered and located in remote areas. Often, just a few soldiers armed only with individual weapons were posted in isolated areas where they could not get assistance from other troops if attacked by the enemy.

In the meantime, a power struggle broke out between Christians and Muslims within Jebha. The Christian faction, also known as the Highlanders, split from the Muslims and set up a political organization that came to be known as the separatists (the Eritrean Liberation Front). Shortly after its establishment, separatists organized a fighting force of its own, and leaving the low-level area to Jebha, moved to the highland. Although Jebha and the separatists held different zones, they fought their common enemy together for a limited period. But, eventually, the conflict between Jebha and separatists led to an intense fight, which ended with the triumph of the separatists. After their defeat, Jebha abandoned its armed struggle.

Following this, the separatists could concentrate on their goal of liberating Eritrea. They amassed a vast number of both light and heavy weapons from their supporters and by capturing them on the battlefield. They recruited Eritrean youths to their side, both voluntarily and by force, significantly augmenting their strength. They organized their troops into divisions to make command and control simpler. The separatist's mode of operation changed from guerrilla to regular warfare.

The separatists destroyed roads and bridges leading to and from Asmara with explosives so that Ethiopian detachments could not quickly get logistical support or reinforcements. They ambushed convoys supplying food. It, therefore, became necessary to protect convoys with heavily armed troops supported by heavy weapons and tanks.

The Ethiopian troops protecting the convoys, supported by fighter jets, made efforts to reopen the closed roads. But the separatists were indomitable. It then became necessary to supply food to our forces by airdrop. With the roads blocked or unusable, the separatists encircled our troops, forcing them into trenches. With mortars and machine guns on surrounding high grounds, they shelled and fired on our soldiers by night and hid from the fighter jets during the day. The separatists were continually feinting attacks in numerous places so that one detachment could not come to the aid of the others.

The first detachment of the Ethiopian army to be annihilated had been assigned to Karora. It was part of the 15th Battalion, under the 3rd Brigade of the 1st Division, sent from Addis to Eritrea to reinforce the 2nd Division.

The 3rd Infantry Brigade of the 1st Division, commanded by Colonel Kinfe Gabriel Dinku entered Eritrea and set up its headquarters at Keren. They were tasked with putting up a defense together with the 16th Infantry Battalion at Afabet and 15th Battalion at Nakfa.

The 15th Infantry Battalion, headquartered at Nakfa, was to send one company to Karora and take Keren village under its control. Karora is a small border village at the northern tip of Eritrea, located between Sudan and the Red Sea. Contraband traders crossing the Red

Sea in small boats from the Middle East (Saudi Arabia and Yemen), make use of the port near Karora village to transfer their goods to the central part of the country. One company was stationed at Karora to stop the entry of weapons and ammunition through the port nearby. The heat at Karora is as high as 40 degrees Celsius. When the sun heats the sand, and the wind blows it, it feels like fire on one's face or body, as early as 0800 hours. The troops were not able to build strong defensive positions because they concentrated so much on coping with the unbearable heat and the chronic shortage of water.

As a result, the company had to rely on foxholes, which gave less protection. Separatists, positioning their mortars and heavy machine guns on the high ground, had been shelling their positions day and night for quite some time. The mortar rounds injured many soldiers, and the morale of the soldiers was very low.

On January 7th, 1977, separatists launched an attack on the company position at dawn. The men resisted and fought the attackers for hours, but when a small number of the troops decided they could endure no longer, they scattered, with some of the soldiers fleeing to Sudan.

A few determined and heroic troops fought against separatists to the last man, killing a great many enemy troops and dying with honor. Several others gave themselves up to the separatists.

After eliminating our forces, the separatists took control of Karora. When the 15th Battalion forces heard the news of the elimination of the Karora company, they were both concerned and angry. When the 15th Battalion established its headquarters at Nakfa, the battalion did not select suitable ground. It was an area not easily defended. The former regional administrative office was selected as the headquarters for the force to make use of the houses as shelters and to have the water well close.

The area selected as the headquarters of the 15th Battalion was in the southern part of Nakfa in a low area shaped like a bowl. The surrounding land to the south and east consisted of fields with several hills suitable for positioning machine guns. The 15th Battalion did not have enough men to cover all the ground and had to choose whether to protect the water and shelter or occupy the strategic high grounds. The 15th Battalion chose the shelter and water, leaving the militarily advantageous high ground unoccupied. When the separatists entered Nakfa from Karora in brigade strength, they immediately took control of all the high ground, constructed defensive positions, and stayed there.

The separatists' biggest problem was the fighter jets. To avoid being attacked by the fighter jets, they dug their positions during the night. To prevent their trenches from being spotted from the air, they camouflaged them, and at daybreak would leave and return to the valleys of Nakfa mountain and remain hidden there.

The separatists were skilled at digging trenches. They dug strong trenches that could withstand aerial bombardments, and their strong position enabled them to control the movement of the 15th Battalion. When work on their trenches was completed, they would place one or two snipers to cover each of our positions, preventing soldiers from moving in or out. They set up mortars and heavy machine guns and kept up a ceaseless bombardment. The 15th Battalion, while resisting the attacks, radioed an SOS to the high command and the Derg, The Derg sent a battalion-sized force to Nanno, 10 to 15 kilometers east of Nakfa by helicopters. But the separatists had prior warning of the landing of the battalion at Nanno. They placed heavy machine guns on the high grounds and waited there for the force to arrive. The reinforcement battalion was forced to retreat. The Derg came to realize that to rescue the 15th Battalion it was necessary to prepare a strong Brigade force supported with tanks and artillery. But it would take time to organize the required men and weapons. There was no time, and action was needed before the 15th Battalion was annihilated.

The only alternative was to drop an airborne unit in by parachute. This operation became the Nakfa Campaign discussed earlier, in Chapter 8-15.

They also planned to send a mechanized brigade into Nakfa within four days. While the Betis Brigade's relief attempt was briefly mentioned in the previous chapters, I feel a better explanation of the operation is needed, and I have provided it here.

The mechanized unit, named the Betis Brigade, completed the

preparations for its advance toward Nakfa. Col. Lemesa was recalled from retirement and was assigned as its commander with a new rank of Brigadier General. The advancing relief column met no opposition from Asmara to Keren. Upon its arrival at Keren, however, the troops refused to continue on the mission unless they were equipped with new weapons, gear, and shoes.

They got what they asked for. However, the soldiers were still unwilling to undertake the mission, and there was a lengthy delay. This delay of our forces at Keren created an opportunity for the separatists to prepare a strong defense on the road to Messalit. The enemy heavily mined the roads leading to Messalit and placed machine guns on all the high ground. When the relief column finally started from Keren and continued its journey to Nakfa, the enemy opened fire as they were crossing Tebabit. The road leading from Tebabit to Mesalit offered no cover or concealment. It lay in an open plain. It is also the only way to get to Nakfa from Keren, especially for tanks and heavy vehicles. If the enemy held Messalit, it would be impossible to get to Afabet without sustaining significant losses.

To fool our attacking force, the separatists constructed both real and fake trenches at the top and middle of the hills, but their primary defensive positions were located at the foot of these hills.

The jets, artillery, and tanks wasted their ammunition on the fake trenches. The enemy was untouched. The advancing infantry, making their way toward the enemy positions, was looking for the enemy in the middle of the hill. However, the separatists unexpectedly rose up as if from beneath their very feet to attack.

As the advancing Ethiopian brigade faltered, the separatists switched to offense. Unable to withstand the assault of separatists, our force made an uncoordinated retreat, abandoning considerable property to the separatists before making it back to Keren. After compelling the Betis brigade to enter Keren, the separatists, to further tighten the noose at Nakfa, brought along two additional brigade forces. On March 22nd, 1977, Nakfa fell into the hands of separatists. On March 28th, 1977, it put Afabet under its control.

When helicopters carrying reinforcements took off from Asmara and flew to Afabet, the radio communication between the Ethiopian units was interrupted. The Ethiopian flag was not flying over Afabet. The mission was canceled, and the helicopters carrying with the reinforcements returned to Asmara. The separatists went on to take control of Tesenei on May 6th, 1977 and Dekemehari on July 2nd, 1977.

Over 4,000 well-equipped Ethiopian soldiers supported by tanks were placed on guard around Keren. The separatists launched their attack on Keren. The Ethiopian troops at Keren, after a battle lasting less than an hour, surrendered to the separatists taking with them their heavy guns, tanks, military vehicles, and ammunition in July 1977. Many believed that the Ethiopian troops at Keren surrendered to the enemy without fighting because the Eritrean created such a distrust of the government among our soldiers in Eritrea. The seizure of Keren by the enemy was a severe blow to the Ethiopian military and their self-confidence.

The separatists, in the tanks and military vehicles they had captured at Keren, and pulling the heavy guns behind them, headed straight towards Asmara. They encircled Asmara and began a continuous bombardment of the city. Asmara, Massawa, and the rest of the Eritrean region fell under separatist control. Only Barentu and Adikeyeh had not fallen, but the separatists had encircled the two cities.

Separatists started to establish political, administrative, judicial, health, educational, and financial offices and began to administer the areas they put under their control. They recruited the youth and sent them to Sawaforto for military training. They were issuing pass-papers through their agents, which was required when moving from one area to another. The separatists established schools and started teaching children as well as adults. They set up clinics and provided medical treatment.

CHAPTER 19
The Closure of the Addis Ababa to Asmara Road

THE SEPARATISTS SKETCHED out a plan to put Eritrea under their control. Part of their plan was to close the following two major roads:

- The Addis Ababa to Asmara road, which passes through Mekele, Adigrat, and Zalambessa.
- The Addis Ababa to Adwa road, which passes through Gondar crossing the Mereb river.

Since the roads were used to transport food and supplies from Addis to troops stationed in Tigray and Eritrea, the separatists wanted to close the roads to create a shortage of food. Previously, a convoy could carry food from Addis Ababa to the troops stationed in Adigrat, Adwa, and other parts of Tigray, and to troops in Eritrea and then move towards Asmara without experiencing an attack.

But as the situation worsened, it became very costly to get the convoy safely to Asmara. When the losses became too high, the food supply was arranged to be taken from Addis to Assab by road, and from Assab to Massawa by ship with the protection of fighter jets and escorting gun boats. Combat engineers continuously cleared the road of mines and explosives while troops guarded the roads from strategic positions on both sides. This was needed to ensure the food trucks accompanied by tanks, heavy guns, and fighter jets could get from Massawa to Asmara.

On occasion, the separatist's attacks on the convoys were successful. The separatists would carry away the food in the trucks when they could and when that was not possible, they used incendiary bullets and explosives to burn the trucks and the food within them to ashes. They continued to fight hard to prevent food and other supplies from reaching Asmara. The damage they caused was so severe that it became necessary to close the road from Massawa to Asmara. The closure was a major defeat to the Ethiopian forces, and a big win for the separatists. The only alternative left for the Derg to get food and supplies to troops in Eritrea was by air drop. The Air Force, with the number of planes it had, was not able to cope with the transportation of the required food and other supplies. Since it was a question of sovereignty, Ethiopian Airlines also helped in the transportation of food and supplies, troops, and ammunition. The Air Force and Ethiopian Airlines jointly fulfilled their mission, but this created a great strain on Ethiopia's economy.

CHAPTER 20
The Eritrean Counter Attacks

A NEW COMMAND WAS established in 1977 in Mekele to direct the liberation of Eritrea from the separatists. The coordinators of the command post were:

- Colonel Tesfaye Gebrekidan
- Colonel Haile Giorghis Habtemariam
- Major Gebrekristos Buli
- Major Negash Dubale

Several Soviet officers were assigned as advisers to the command.

The entire Ethiopian Army command moved to Mekele. The new command set aside enough supplies, including food, weapons, and ammunition for a six-month long engagement. Medical personnel and supplies set aside for the upcoming operations and surgical hospitals were established in both Mekele and Asmara. When all the logistics were in place, soldiers were withdrawn from the eastern and southern fronts, and were moved to Mekele and Asmara. The troops in Mekele were organized into five task forces; 501, 502, 503-A, 503-B, and 503-C.

The Massawa troops were designated as the 505 Corps task force and the Asmara troops were designated task forces 506-A and 506-B. The combined force included 86,722 men, 155 tanks, 180 heavy weapons, and 24 BM 21 rocket launchers,

A task force is established to carry out a certain task on a temporary basis, and when the task is completed, the task force is to be dismantled. Therefore, the structure of an individual task force as far as the number of men and type of equipment varies depending on the type of task to be carried out. In a task force, there may be three or four infantry brigades, or perhaps three or four infantry divisions, along with any needed support units such as armor, field artillery, etc.

The organization of the new force was complete but there were not enough high rank officers with fighting experience to command the newly formed task forces. Therefore, the Derg recalled all high-ranking officers who were training abroad to address this shortage. I was among twenty high-ranking officers who were studying to become brigade commander and staff at the Frunze Academy in Moscow. In August 1978, the Ethiopian Ambassador to the Soviet Union came to the Academy and informed us that, an order from the Derg was posted and, we needed to stop our studies and return to Addis.

The next day, we boarded an Antonov 12 to fly back to Addis Ababa. We flew out at midnight, stopped in Cyprus for a refill and landed at Bole International Airport at 0600 hours. We were welcomed by the then administrative officer of ground forces, Colonel Getachew Gedamu. As soon as we got into the terminal, he took us to the reception hall and told us that those of us with families around Addis Ababa, Debre Zeyit, and Nazret, would be provided with transport so that they could spend the night with their families. However, we needed to report back to the unit's assembly at 0900 hours with all necessary equipment.

The high military officers who were abruptly ordered to halt their training in the Soviet Union and return immediately to Ethiopia, 1978. Author: back row, 4th from left.

Pictured with the school commanding officers in the Soviet Union training academy in 1978. Author: 2nd row, 1st right.

I spent the night with my family in Debre Zeyit, and the next morning returned to Addis. Many of the officers who returned with me from Moscow were waiting for the arrival of Colonel Getachew Gedamu in the assembly hall.

A couple of them came to me and said, "You have already done your part, you should take a break on the current war in Eritrea, it's now our turn to fight."

Most of them knew that I fought both in the eastern and northern battlefronts and that I was wounded twice.

Colonel Getachew came in, holding a sheet of paper in his hand.

"Officers listen to me, I'm going to call out names," he said. Everybody was listening. "I'm going to call out names of officers who will be flying to Asmara today. When you hear your name, please proceed with your luggage and go straight to the bus to Bole International Airport."

My name was the first one called. He went on to call out ten additional names.

"The rest of you, until I receive another directive from the Derg, will report to me every day, at 0900 and 1600."

Those whose names were not called out were persons related to or friends of the Derg members. Those of us who had no connections always became the first fighters during all of Ethiopia's wars. We shuttled back and forth between the eastern and northern battlefronts, or wherever a shot was heard. At least, that sure was how it seemed.

Those of us who were told to go to Asmara that day requested Colonel Getachew to give us one day of leave so that we at least present the clothes we bought for our children while we were in the Soviet Union. When we had arrived the previous night, our children were sleeping and when we left home this morning they were not yet awake.

"I cannot give you that permission," Colonel Getachew Gedamu said to us. "I have no authority to do so. The only person who can give you that permission is Major Gebre Kristos Buli. You can approach him if you wish to do so."

Major Gebre Kristos Buli was the Operations Officer and located at the Derg offices, making it difficult to access him in person. However, we could contact him by telephone. As I was the most senior in rank among the team going to Asmara, I spoke by telephone with Major Gebre Kristos on our behalf.

I asked him to give us a one-day leave and explained the reason. He reluctantly agreed and he asked me to pass the telephone to Colonel Getachew. I don't know what words the two exchanged, but I believe Major Gebre Kristos scolded Colonel Getachew. His face suddenly shrunk.

"So! You had me scolded! Thanks for that! Now you can go, and come back tomorrow," he said to me. We spent the night with our families, and the next day we flew directly to Asmara on a DC -6.

We thought we would go straight into a battle as soon as we arrived in Asmara, considering the haste in which we had traveled. However, we spent the next fifteen days doing nothing but playing card games. On the fifteenth day, the commander of Asmara area military force, General Merid Negussie, called us to his office, and briefed us on the situation in Eritrea. He told us that the trenches prepared by separatists around Asmara were constructed with strong railway steel that was indestructible, even by air, and that the separatist were bragging about this, saying it would take ten years for the Derg troops to break through them.

Although it might not take ten years to break through their strong defensive position, it would certainly not be an easy task. The cost would be high. He explained to us that the preparations for the assault would be completed soon.

I was given command of the 506th B task force. The 506th B task force is composed of three infantry brigades, one tank battalion, one 122mm artillery battalion, and one combat engineer company. The 506th task force was mostly composed of militia. The militia members had not been given enough training because there were not enough trainers for the number of militia trainees. At times, the ratio of trainee to trainer was more than hundred to one. Some of the trainers did not have enough experience in the field and had not even received instructor training. Some of the trainers were recruited from among the militia themselves.

Training had to be given in ten different ethnic languages through interpreters since there was no common language that could be used for all the militia. Even then, those who understood the subject matter of the training were less in number than those who did not. Some of the trainees were just imitating what they saw others doing without understanding the verbal instruction. When one sees another person sitting, he sits down. When one sees the person next to him run, he runs. When the person beside him lays down, he lays down. When asked why, he would say that he saw the other person doing it.

There was a great shortage of equipment. And, since it was the rainy season, it often rained all day long, which had an adverse effect on the training. The militia were recruited mainly by force. They didn't want to be here and made all kinds of excuses to avoid being on the training ground. Even those who did attend the training were not attentive. They would look down when the instructor was teaching and fiddle with twigs. A military force must be disciplined and submit itself to the rules and regulations of its commanders. Training is more effective when it is demonstrated in action. A commander has more responsibility than any other person in making sure his troops are well-trained and well-prepared. I called on all the brigade commanders and department officer regarding training and discipline, so that we may have a common understanding of how to carry out the tasks ahead of us in such a short time, so we could be victorious. We needed to get out of our offices, ignore our rank, wear our working uniforms, and join in the training with our troops to show them that we to, followed the rules and regulations that we asked them to follow.

The commanders and officers did everything together with the trainees. This increased the trainee's confidence and spirit, and the trust between the troops and commanders. We simulated war with tanks and heavy gun crews and trained them on their tasks and responsibilities. The troops showed a remarkable change in a short period of time. After only one month of training the militia men, who were earlier considered incompetent, quickly developed their fighting skills to the level of regular troops. The training continued until the day we received the order to go to the battlefield.

Then, the militia men started singing, "I shall shoot and kill. I shall vanquish!"

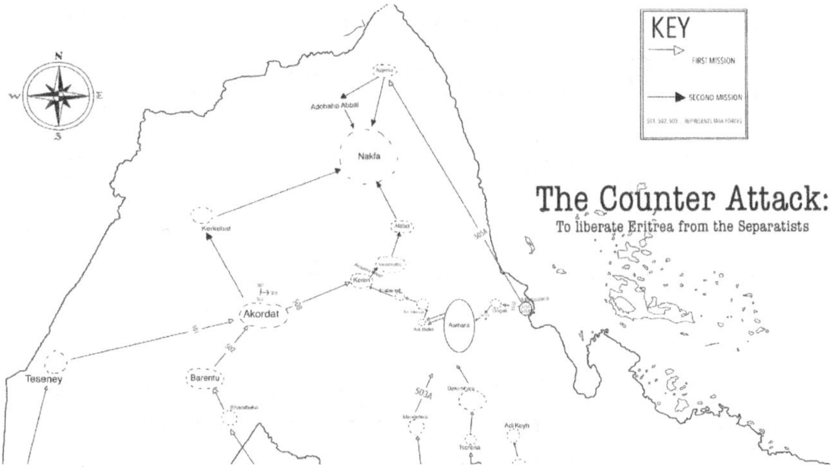

Map 5: Liberation of Eritrea from the Separatists.

CHAPTER 21

The Breakthrough of the Trenches Around Asmara

All the renowned artists of Ethiopia arrived in Asmara to motivate the armed forces before the commencement of the big battle. Before the soldiers left for battle, the legendary singer, Tilahun Gessesse, chanted his most popular morale booster,

"Chase the enemy! Push on!"

The troop's spirits were instantly lifted, and they shouted back, "We want to fight now, let us go!"

On November 20th, 1978, after the show was over, the committee that was tasked with coordinating the war in Eritrea summoned all commanding officers involved in the Asmara operation to a meeting. The committee briefed us on the missions of the different commanders.

The 501st task force under the command of Colonel Regassa Djima as its first mission, shall start from Gondar, cross the Tekeze river, reach Humera, clear the separatists from Humera and enter Humajer. They will shift their direction towards the east and close the exit door of the separatists trying to flee from other fronts to Sudan. Hit them hard and proceed to Tesenei. After reaching Akordat, it shall reorganize itself and be ready for the next mission.

The 502nd task force under the command of Colonel Asrat Birru, will start their first mission from Mekele, pass through Temben, Adwa, Shire, Shiraro and cross Badime and Shembeko, and enter Barentu. After taking control of Barentu, they shall proceed to Akordat. On reaching Akordat, they shall reorganize and be ready for the next mission.

503 Task Force coordinated by Colonel Kassa Gebremariam, is subdivided into A, B, and C:

- The 503-A task force under the command of Colonel Aytenew Belay, shall start from Mekele, pass through Temben, enter Adowa, cross the Mereb river and take Adikuala and Mendefera under control. They would reorganize themselves at Mendefera and wait for their next mission.
- The 503-B sub-task force, commanded by Lt. Colonel Tariku Laine, as their first mission, shall start enter from Inticho and proceed to Tsorena via Inticho, assault and take Tsorena under their control, then enter Dekemehari. They will re-organize themselves at Dekemehari and wait for their next mission.
- The 503-C sub-task force, under the command of Lt. Colonel Seifu Wolde, as their first mission, shall start from Adigrat, then assault and place Zalambessa under its control. It shall then proceed to take Adikeyih, Senafe, Ginda under its control. They shall establish contact and connect with the 505th task force and get ready for their next mission.

The 505th task force, under the command of Colonel Abera Abebe, as their first mission, shall first liberate Massawa town, take Dogali, Maiatal, Gatela and Dengolo, Ginda and Nefasit, under their control, open the road between Massawa and Asmara, and ensure that it is open for use. It will reorganize at Nefasit and be ready for its next mission.

The 506th task force sub-divided into 506-A and 506-B and coordinated by Brigadier General Merid Negussie.

- The 506- A sub-task force shall be under the command of Colonel Hussein Ahmed.
- The 506-B sub-task force was set to be under my command, Lt. Colonel Tesfaye Habtemariam.
- The two sub-task forces, 506 A and B, will jointly break the trenches around Asmara, and chase separatists until they reach Adibidil and then assault and take Aditeklezi under our control. They would reorganize themselves there and get ready for their next mission.
- Then 506-B task force shall march directly from Adibeyil to Shemangus Tatai, take Shemangus Tatai under its control and open the Asmara-Keren road to road traffic. It will then prepare for its next mission.

The operation would begin on November 21st, 1978 at 0530 hours.

The Ethiopian forces were determined to break the Asmara trenches at all costs, knowing that if they did break through the trenches then eighty percent of the Eritrean war would be over. They also knew that breaking through would come at a high cost.

The infantry was waiting the order to attack. In the meantime, we kept busy cleaning weapons, loading magazines with bullets, checking and rechecking equipment, and send out scouts to monitor the movements of the separatists.

Our combat engineers were waiting for dark to begin removing the anti-personnel mines and booby-traps that were in our path. Over time, these had been placed by both by the enemy and the Ethiopian forces.

Over one thousand heavy guns and mortars were in place behind the infantry to carry out the preliminary bombardment from their concealed positions. Each heavy gun was supplied with two loads of powder and was waiting to begin firing. Fighter jets fully fueled and loaded with missiles and ammunition awaited take off to provide air support.

Tank and armored car crews had taken their positions behind

the infantry to launch an assault together with the infantry. All units, having completed their preparations, waited for the attack to begin.

On November 21st, 1979, at exactly 0445 hours, the artillery crew signaled to herald the beginning of the preparatory bombardment. Other heavy guns, following the first one, shelled the enemy position continually for forty-five minutes, concentrating fire on the point selected for the planned breakthrough. The area was pounded, covered with the flash and smoke of the heavy gun's exploding shells. One-minute before 0530 hours, the artillery crew turned its firing towards the second trench.

The troops of task force 506-B, taking advantage of the smoke from the bombardment, threw hand grenades into the first trench. Then, with fixed bayonets, they charged into the trenches. Exchanging bayonet thrusts with their enemies, they fought hard and succeeding in creating an opening in the enemies' defensive line… and finally broke through.

Unable to withstand the assault of the Ethiopian forces, the separatists fled from their trenches and began to flee towards the west. With support from the air, and from artillery, our men gave chase and continued to scatter them.

The trenches around Asmara, the pride and glory of the separatists, were smashed and overtaken by our forces in less than an hour. And with far greater loss of life than was expected. When the fight moved from the trenches into the open field, the separatists were in full flight and our men pursued them hard.

Before we reached Adibidel, the separatists command and control had disintegrated, and they fled towards Elabered and Keren in a scattered and cowardly manner. When the Ethiopian forces reached Adibidel, 506B task force headed north towards Aditeklezi, to take the Asmara to Keren road under control. It took us some time to reach the other side because of the jagged cliffs. We had to spend a night on the way, before we could reach our destination because it got too dark. The battle maps we had did not coincide with the landmarks on the ground. The maps were from the Italian occupation time and they were outdated.

The battle plan map showed a level plain, but we had cliffs before us instead. The infantry eventually managed to scale the cliffs, but the supply trucks had to make a U-turn to Asmara together with task force 506-A.

At dawn, on November 22nd, 1978, the fighting began and lasted for the entire day. We were able to take control of the entire area at about 1600 hours. We took the high ground and spent the night there. Vehicles loaded with food and clothing had to wait in Asmara until the engineers verified that the road was free of mines.

After a period of forty-eight hours without food and water, we found and ate raw maize and drank stinking water full of worms from a pool near where we spent the second night. We could not enter Aditeklezi on the same day because of the slow down caused by the difficult terrain and had to spend a night in a temporary defensive position.

On November 22nd, 1978, moved forward, we cleared the area, and seized control of Aditeklezi at 1500 hours.

The 505th task force started out from Massawa on November 21st, 1978, and promptly reached Nefasit and took it under control. Shortly after, the Massawa-Asmara road was re-opened.

Taskforce 502 moved through Badime and Shenbeko, joining with the 507 sub-task force near Barentu, and then quickly took Barentu and its surrounding area under control and then moved to Akordat to join with the 501.

The 502nd task force crossed the Tekeze river and entered Humera.

The 503-A, the 503- B, and the 503-C task forces could not cross to Eritrea from Tigray because the Mereb river had overflowed its banks and the bridge had washed away. The forces had to wait until the river subsided.

The coordinator of 506- A and the 506-B task forces, Brigadier General Merid Negussie, seeing that Aditeklezi was now under our control, moved from Asmara to Aditeklezi to establish a forward-command-post and better control combat operations.

On November 22nd, 1978, at 2200 hours, I received a radio message from the operations coordinator, Brigadier General Merid Negussie, urgently instructing me to come to the command post. When I arrived, General Negussie seemed to be in low spirits. I approached him, saluted, and sat down.

"Colonel Tesfaye, we are in a mess," he said.

"What happened?" I asked.

"Radio communication with Colonel Hussein Ahmed has been lost," he replied.

"What do you think happened to Colonel Hussein?" I asked.

"Gebre Kristos had flown in with his helicopter and ordered me to have the 506-A sub-task force move their tanks immediately and to spend the night in Elabered. I told him that I can't follow his order because it contradicted the battle order I gave earlier and because it was too late. Kristos replied that he thought I was afraid and if so, he would have me replaced. Then he flew back to Asmara. I was very annoyed, but I ordered Colonel Hussein to head for Elabered and spend the night there."

Colonel Hussein had reported that he'd arrived at Elabered and communication were lost shortly after, about 2100 hours, just a few minutes after he told me that the battle had begun.

"Now, what should we do?" he asked me. Then he went on, "your task force should move quickly and give him support."

"I am not familiar with the terrains of Elabered, but it is common knowledge that Elabered's terrain is difficult and mountainous and the troops have no experience fighting at night," I replied. "If we move out now, we will fall into the hands of the enemy before we can do anything. Instead of making the same mistake as Colonel's Hussein's force, it is better if we start our mission at dawn."

We agreed, and I returned to my command-post. On November 23rd,1978 at 0400 hours, my troops and I started moving to Elabered. Five kilometers short of Elabered, the delaying force of the separatists opened fire at us.

We moved to control the high ground on both sides of the road and began a counter-offensive. We overran the separatists' delaying force and proceeded forward. Elabered is surrounded by moderately high mountains. The Asmara to Keren road passes through Elabered. There is only one opening that serves as entry and exit for vehicles. Elabered's lowlands have fertile soil carried down by the runoff from the mountains. The dam that collected the rain and runoff was used to irrigate crops of fruit and vegetables. Looking down at Elabered, it was covered with green leaves of orchards and groves of fruit trees such as oranges and mandarins. Elabered was also known for its pasteurized, bottled milk, which was distributed all over Ethiopia. The side that controlled the high grounds of Elabered had a better chance of victory.

The separatists, forced out of their trenches, had fled in a disorganized manner. Some of these troops were hiding in Elabered while their leaders attempted to get reorganized. The separatists could not believe their eyes when they suddenly saw tanks and military vehicles of 506-A task force quickly approaching. They were not used to seeing Ethiopian forces marching after 1700 hours.

The scouts of the separatists warned that enemy tanks were approaching Elabered, and the leaders ordered their men to prepare for battle. They were ordered to wait until the task force came through the gap, then close the gap with anti-tank rocket launchers, and stay hidden until all the Ethiopian forces had entered Elabered. Unaware of the situation, the 506-A task force entered Elabered without taking any fire. Once in the city, they dismounted from their tanks and other vehicles, and began seeking shelter under trees. They had spent the whole day traveling and fighting and wanted to eat and rest. Putting down their weapons, many went to look for water. Others were collecting twigs to start fire for making tea and to cook a meal.

Once the separatists ensured that the men of 506-A task force were in the city, and had lowered their guard, they attacked with a hail of hand grenades, mortars, and light weapons fire. The Ethiopian forces were taken completely by surprise. In a desperate fight, the soldiers of 506-A fought hand-to-hand with the separatists using bayonets and throwing hand grenades. Soldiers on both sides fell to the ground side by side. Radio communication was lost. There was no longer any cohesive command. Most of the tanks and military vehicles were destroyed by the separatists. The sky was filled with black smoke. Heavy gun and tank shells were exploding on their own. It was not possible to differentiate between the shots fired by the Ethiopian forces and those fired by the separatists. Bodies of the fallen troops were so badly burnt that it was impossible to distinguish those of the Ethiopian soldiers from the separatists.

Fighter jets could not tell who they were bombing since the Ethiopian forces and the enemy forces were mingled. Many of the separatists were wearing the uniforms of the fallen Ethiopian soldiers. The jet's bombs killed men on both sides.

When 506-B task force arrived at Elabered, I ordered the advance brigade to send one battalion to take the high grounds on the right and reserve two other battalions. The second brigade was ordered to take the high ground on the left side. I held the 3rd brigade as a reserve, ordering them to provide fire support to the two brigades as required. I gave the brigade commander an order not engage their reserve troops without notifying me first. Too often, Brigade commanders would commit their reserve troops into battle too soon and waste their fighting force unnecessarily. I went with the advance brigade to the right side. I wanted to be close to the action to better command the two brigades during the action.

Together with the surviving troops of the 506-A task force who had since re-grouped, we fought, defending ourselves with fierce determination. By roughly 0100 hours, 506-A task force, was in disarray. Abandoning their positions, the retreated. Following 506-A task force, 506-B task force also retreated. I looked around and my radio operator and my bodyguards were nowhere to be found. I was all alone.

Mortar shells were flying and exploding everywhere. Fragments from mortar shells whizzed past my ears. I knew I could be killed at any moment. I decided to keep my faith and courage, and to keep moving closer to my men, at least until I was struck down.

After a short distance, I saw our tank commander, Major Gezmu Bezawork. He was an old friend of mine. He had been severely wounded by a mortar round and had fallen near his tank. Both his legs had been badly mangled in two or three places directly below his buttocks. As soon as he saw me, he called out,

"Colonel Tesfaye, I beg you in your children's name, finish me off!"

He had lost a lot of blood.

"Be Strong! If we are meant to die, we will do so together, or we will survive together. I will not use any bullets on you. I won't leave you here, either." I carried him on my back. Together, with bullets whizzing and shells exploding all around us, we started moving towards our men.

When I picked him up, I was at a loss about which part of him to hold onto. Parts of his body were barely attached, looking like they were about to fall off. We somehow reached our lines. I put him down under a tree, instructed some men to watch over him. Then I went to find a vehicle to drive him to where he could get airlifted by a helicopter that would take him to Asmara. Major Gezmu Bezawork is still alive today, and I feel great satisfaction that I brought him to safety and did not end his life as he asked me to.

The troops had left their temporary defensive positions in the evening and retreated under the cover of darkness, leaving us behind. We could not stop them. When told to stop in the name of the Lord, in the name of the Ethiopian flag, they threatened to shoot us. The 506- B task force, mostly the reserve brigade, retreated, and joined with the troops that had already retreated. The only ones remaining were the officers, including me. The main battlefront was wide open. Considering the seriousness of the situation, Brigadier General Merid Negussie gathered together all the officers and ordered that we all, himself included, hold the defensive position ourselves.

"If we are going to combat in the trenches, who is going to lead the troops? If we die who is going to give to lead them?" We suggested that it would be better to gather our troops and cover the opening.

"Who is going to cover the opening until you go and come back with the troops you gather?" he asked.

His concern was valid. If the separatists move towards Asmara with their tanks, there would be no force to stop them from entering Asmara.

To quiet the concern of Brigadier General Merid Negussie, I said, "If my reserve battalion is still where I positioned them, with their tanks and heavy guns, I will send one company from that battalion to cover the gap."

"Fine, Colonel Tesfaye, send a company to cover the gap. The rest of you officers, go back to your units and regroup, get them back under control." Then, he let us go.

I took one company from the tank and heavy gun units and positioned them to cover the gap. Then I returned to my troops. The separatists were so exhausted that they could not launch an attack that night. But we used the night to regroup our troops and re-establish command and control. The 506-A task force was given the task of taking the area south of Elabered under its control, while the 506-B task force was given the task of taking the area north of Elabered under its control.

On November 24th, 1978 at 0530 hours, we launched our assault supported by fire from our heavy guns and fighter jets. 506-B task force moved to the north and fighting an intensive battle, began to take control of the assigned area.

We coordinated our assaults with fighter jets and the heavy gun crews. Our tanks fired straight into the separatists' defensive works, demolishing them. The infantry engaged in hand-to-hand combat, many were killed. We were finally able to take the area under our control.

Once we held the high ground, the separatists were forced to abandon Elabered and retreat towards Keren and the western desert. The time was around 1500 hours when the separatists began to flee towards Keren. There was enough time before nightfall to pursue the separatists, forcing them to enter Keren. Then we got orders from the high command to stay in Elabered, so we returned to our earlier positions and spent the night there.

It was horrific to see what was left on the grounds after the battle. Severed limb sand heads were scattered everywhere, bloody and filled with guts. Bodies with crushed skulls, eviscerated entrails, dead bodies piled one on top of the other, bodies charred from fire and partially eaten by wild animals and vultures were strewn all over the field.

Equipment worth millions had been destroyed, burnt, and

smashed. Thousands of human lives were lost. Elabered, once covered with greenery, was now a barren wilderness with its grass and trees all scorched. It was not possible to distinguish friend from foe and all the bodies were buried in mass graves dug with graders and bulldozers. Over one hundred bodies were placed into each of the graves.

The 502nd Task Force from Barentu and the 501st Task Force from Tessenei entered Akordat. The two Task Forces merged into one Task Force at Akordat and was renamed the 508th Task Force. The 508th, led by Colonel Asrat Biru, followed the 3rd Division to Keren, leaving the 506th Task Force to control Elabered.

The separatists assigned just a small force to delay the movement of the 508th Task Force from Akordat to Keren and sent their main force to Messalit to repair the old trenches they had dug in 1977 and to construct more trenches. They also mined all the approaching roads with anti-tank and anti-personnel explosives.

The 508th Task Force, upon arrival at the entrance to Keren, faced stiff resistance from the separatists' delaying force. The battle was prolonged because the 3rd Army Division could not smash through the separatists and enter Keren. The 506th Task Force was given the task of marching from Elabered to Keren and help the 508th Task Force overcome the separatists. The 506-B Sub- Task Force started off from Elabered and took Keren under its control on November 25, 1978. To avoid getting caught in a trap, the separatists delaying force retreated as soon as they saw that the 506-B Task Force had entered Keren.

The 506-B task force and the 508th task force were given 4 days of rest after entering Keren. Then, they continued to take control of Afabet and Nakfa via Messalit, by using the motor road that passes through the Messalit hills since that was the best road for vehicles and for infantrymen.

The Ethiopian forces had to cross an open plain without any cover or concealment to get to Messalit from Keren. From the hills, as our men crossed the plain, the separatists fired on them using heavy guns and mortars to delay their passage.

The 506th Task Force entering Keren Town in triumph, 1978. Author: Front Left

The range of mountains to the east that ran from north to south became a great obstacle for fighter jets trying to fly at low altitudes and strike their targets accurately.

The Ethiopian forces launched repeated assaults to take Messalit under their control. The repeated attacks began to take its toll. The separatists began to attack when they saw that our men were tired. The Ethiopian forces fled to Keren leaving behind many heavy guns and tanks.

After the 506th and the 508th Task Forces entered Keren, the 506th Task Force was merged with the 508th Task Force and named the 508th Task Force. After the 508th Task Force reorganized itself at Keren, Brigadier General Asrat Biru was assigned as the commander of the Task Force. The 508th Task Force launched its second assault on Messalit with greater strength and was able to break through Messalit and put Afabet under its control, later arriving at Nakfa gate.

After their defeat at Messalit, the separatists moved straight to the trenches they had dug in Nakfa in 1977. The Ethiopian forces had to take a defensive position on the high ground at the gate of Nakfa and were forced to stay there since they could not cross over the Nakfa ridge. Each command was given a permanent defensive area at Nakfa gate and was supplied with heavy guns, tanks, anti-aircraft and anti-tank weapons. We began a coordinated defense of our assigned areas of responsibility.

After the 506th Task Force merged with the 508th Task Force and was reorganizing at Keren, I was assigned as administrator of Keren and of the province. There was no civil administration at the time, so I was also assigned to take care of any civil administrative responsibility.

The inhabitants of Keren had no water, power, or telephone service. There had been no banking services for years. Clinics and pharmacies had been closed, and no medical services were available. There was no public transport service between Keren and Asmara. Consumer goods like sugar, coffee, cigarettes, drinks were only available as smuggled contraband brought in on pack animals from Asmara and sold at very high prices. The first step I took to address the people's water problems was for our troops to deliver water to the public using water trucks. Then I contacted the water and power authorities. In a short period of time, pumps for water and generators for electricity were installed in Keren and this problem was solved. All the required human and material resources, both for installation and maintenance were also flown in from Addis.

I arranged for the Commercial Bank of Ethiopia and Telecom to open branch offices in Keren so that residents of Keren could resume their business as quickly as possible. After my request to the Eritrea communication and transport office to re-open the Keren Asmara motor road, buses were assigned to provide transport to the public from Keren to Asmara and vice versa. Traders who earlier had trading licenses were advised to renew their permits and help control the soaring prices of goods by supplying consumer goods to the public at reasonable prices. The price of essential commodities such as sugar, cigarettes, coffee, and drinks were fixed to protect the public from being exploited.

I asked residents of Keren to select elders to represent them in discussions with us about security or other concerns they might have. The troops were given strict instructions to respect the rights of the people. They were told not to touch property of the people without their permission, and to pay market price for the goods they bought, and to refuse bribes.

I did what I could so that Keren residents could conduct their daily lives and business feeling safe and secure, with reliable services and infrastructure. The troops and the residents learned to respect each other and help each other, working and living in harmony.

Being awarded the highest possible medal in the Ethiopian armed forces, the hero medal of Ethiopia, from the president of Ethiopia, 1980. Author: Left

In July 1979, I was told to be in Addis as soon as possible. On September 12, 1980, at Revolution Square and in the presence of Comrade Chairman Mengistu Haile Mariam, Derg members,

Ministers, Ambassadors, Military attaches, and representatives of the Ethiopian military forces, I was awarded the Battlefield Hero Medal of Ethiopia.

On the same day, many other officers and Non-Commissioned Officers (NCO's) were awarded medals in recognition of their service, 1980. Author: Front row, 1st Left

This is me, giving a speech on behalf of heroes at the Award ceremony in 1980 at the Revolution Square in Addis Ababa.

The Provisional Military Government of Socialist Ethiopia

<u>Hero Medal Certificate of Socialist Ethiopia</u>

The medal and emblem awarding commission, with the consent of the Provisional Military Government of Socialist Ethiopia, to award the Hero Medal of Socialist Ethiopia to Lieutenant Colonel Tesfaye Habtemariam. This award is based on the medal and emblem proclamation No. 169/1979 for the unparalleled heroic feat he accomplished on the battlefield in defense of Ethiopia and serving his country with full self-confidence, courage, determination and risking his life.

September 13, 1980

Chairman of medal and emblem according commission.

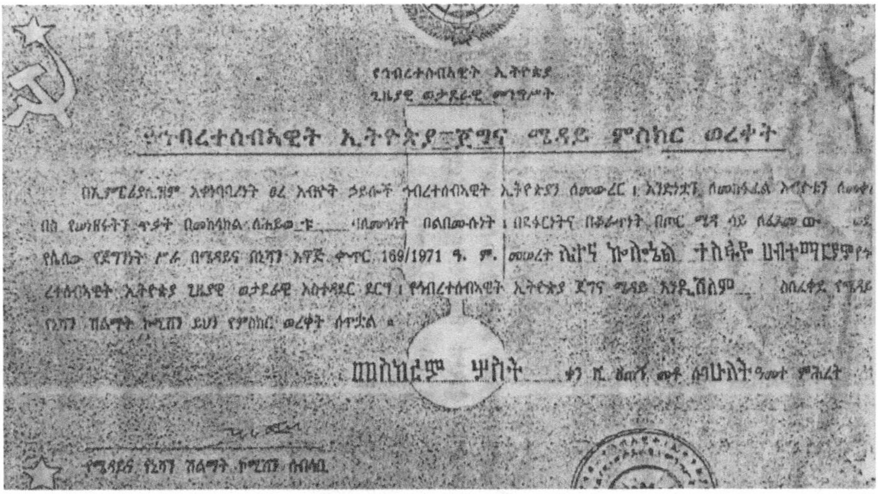

የኅብረተሰብአዊት ኢትዮጵያ
ጊዜያዊ ወታደራዊ መንግሥት

የኅብረተሰብአዊት ኢትዮጵያ ጀግና ሜዳይ ምስክር ወረቀት

The original certificate in Amharic, 1980.

In 1980, when the 18th and 19th Mountaineer Divisions were established, I was assigned as commander of the 18th mountaineer division. After structuring, training, and arming the 18th Mountaineering Division for a few months, I was sent to East Germany for medical treatment to get the fragment in my skull checked. Before I finished my medical treatment, the Ethiopian ambassador in East Germany, Berhane Jembere, came to the hospital and told me he had been instructed by the Derg to inform me that I needed to return to Addis immediately.

On my arrival in Addis, I was given an invitation letter to attend the Inaugural ceremony of the establishment of "Isepako" (the Ethiopian worker party organizing committee). I also found out in the same ceremony that I was one of the 120 central committee members of Isepako. When Isepako was transformed into "Isepa" (the Ethiopian workers party), I was also made one of the members of the central committee.

When the conference establishing Isepako was concluded, I was assigned as deputy training officer of the ground forces. In this same year, I was promoted from Lieutenant Colonel to the rank of Colonel.

CHAPTER 22
Re-establishment of the Airborne Unit

The Airborne Unit started off as a single company when it was first established in 1958. Since then, it has grown to a battalion. The unit also provided parachute jumping and commando training to young officers who had already received training in military science. This greatly increased the battle capacity of Ethiopia's ground forces. The unit was deployed on several missions and recorded outstanding results.

The Derg considered the Airborne Unit one of its strongest supporters during the early days of its administration. During 1975 and 1976, the Derg began to suspect that the unit had been infiltrated with opposition political parties, especially the Ethiopian People's Revolutionary Party. In 1977, the Derg dismantled the Airborne, transferring its members to different departments of the Ethiopian army.

Between 1980 and 1981, it became clear that Ethiopia needed an Airborne Unit. The Derg approached the Soviet Union to explore ways of re-establishing a new airborne brigade. The Soviets offered to establish one airborne brigade in two years. But, demanded they be paid millions in American dollars. They also imposed two additional conditions, all the trainers had to be from the Soviet Union, and all Israeli and American-made training resources had to be replaced with Soviet equipment. This included the disposal of five hundred parachutes (worth $4,000 each) in storage. They also stated that they would not allow any training personnel from Ethiopia or any other country to be involved, the trainers would be brought in from the Soviet Union exclusively. They were confident that Ethiopia had no capacity to establish an airborne brigade without their help, and that no other country would come to assist Ethiopia.

The Derg did not find the Soviet's demand feasible and had to explore less expensive alternatives. As part of that project, I was called to a meeting in the presence of Lieutenant General Tesfaye Gebrekidan, the Minister of Defense, Major General Mesfin Gebrekal from the training department, and Major General Seyoum Mekonnen from the intelligence department.

The Minister of Defense started off by explaining the country's need to establish a new airborne brigade. He went on to explain the offer made by the Soviet Union, but the amount of money they demanded would be beyond our country's means.

"We called you to this meeting to hear your opinion on ways to establish the airborne unit with less cost."

I answered, "Yes, I do have knowledge and experience in airborne and commando training. I have had six months training in fundamental airborne jumping and commando skills in the Debre Zeyit Airborne Training School, six months of Special Forces (Green Beret) and airborne training at Ft. Bragg in North Carolina in the United States, three years as an assistant trainer with the Israelis in establishing a Special Forces group in Dire Dawa. However, one person cannot possibly have all the knowledge or skills required to train an airborne brigade. The training must be carried out by experts trained in different skills, and working in cooperation, helping each other. Also, the training is high risk. There is a great risk and trainees can get hurt, or even killed.

I don't know the state of the training facilities of the former airborne training school in Debre Zeyit since it has been abandoned for years. I don't know where all the trainers are since they have been transferred to different units of the army. Therefore, I ask for one week to fact-find and give you my proposal."

He agreed with my suggestion and the meeting adjourned. I left for Debre Zeyit to inspect the condition of the former airborne training camp.

The training materials were no longer there and what was left was not usable. The living quarters were dilapidated. The parachutes in the warehouse were not in usable condition. The compound had fallen into disuse and had been overtaken by wild life, the ground covered with acacia and many thorny wild plants.

After completing my visit and giving the matter considerable thought, I proposed the following to the Minister of Defense:

- Engineers should be sent to Debre Zeyit to repair the dilapidated houses, tower, swings, ramps, sand berms and simulated airplane.
- A directive should be issued to return the former members of the airborne unit to Debre Zeyit from where they were transferred.
- The Ethiopian Army should supply material for bedding, kitchen utensils, cutlery, and office materials.
- Health department should establish a clinic and assign health personnel, medical equipment and supply medication.
- Communication department should install telephone and radio systems and assign technicians.
- The transportation department should assign trucks, transport vehicles, and ambulances, drivers and mechanics.
- The Army should assign administrative, logistics, and finance personnel.
- Letters should be sent out to regional first secretaries of ISEPA, administrators, the police, directors of schools and to health officers, to assist in finding good recruits.

- The Air Force should provide aircraft for parachute jump training.
- Soviet suppliers of new parachutes should train our staff on how to use the parachutes or we send our people to the Soviet Union for training on how to use them.

The Debre Zeyit Military Camp could accommodate only one battalion, it was not large enough to accommodate a brigade-sized element. I proposed that the infantry and commando training and airborne training be done in an open field in a nearby location. The field could be equipped with tents, water tankers, field stoves, and naphtha stoves, and for all training material to be supplied on time. Only the actual jump training would be done at the Debre Zeyit Military Camp.

The Minister of Defense accepted my proposal and issued an order directing all those concerned departments to give their full cooperation. He stated that the Ministry of Defense will also give me all the necessary support per my recommendations.

"I shall follow the matter closely. You may start your work tomorrow," he said.

Within fifteen days, the former airborne unit members who had been reassigned arrived in Debre Zeyit and immediately started their refresher training. When the trainers completed their training, the following recruitment criteria was set up:

- Recruits should be 18-20-year-old, able-bodied and unmarried.
- Recruits should have at least a 10th grade education or higher.
- Recruits must pass a physical medical exam.
- Recruits must be 1.65 meters to 1.75 meters tall.
- Recruits must weigh 50 to 70 kilograms.
- Recruits cannot be smokers, or users of khat (a plant whose leaves are chewed as a stimulant) or addicted to alcohol.
- Recruits cannot be under litigation or wanted for a crime.
- Recruits must be good at sports, especially soccer and swimming.

- Recruits must be able to speak, write and read Amharic, the language chosen by the Constitution as the official language.
- Recruits cannot be a member of any political party.
- Recruits cannot have had any major surgeries or any fractured bones.
- Recruits must pass airborne training school entry exams and have good grades in behavior in school.
- Recruits must pass visual and hearing test with 100% and be willing to be trained in airborne skills.

All regions with high schools were given a quota to produce a multi-ethnic force. Within one month, 2,600 men were recruited, and sent to the training ground in the Awassa Lake area.

The recruits were organized into 4 battalions and were given a four-month training in basic infantry skills. When the basic infantry skill training was completed, commando and airborne training was given. We improvised sand-barriers, ramp tower and swings. We slept on the ground, under trees, and trained day and night. When the battalion sent to Debre Zeyit completed its jumping exercises, it would go back to Awassa Lake and the next battalion would start its jumping exercise in Debre Zeyit. This way, the training continued in shifts.

The Soviet advisers made several attempts to obstruct the training so that it might not be successful.

"What entitles you to train airborne men? It's better that you stop this before a disaster occurs, unless you want them to get killed," they would say.

"Never mind about that," I would reply. "Just leave my presence and go where I can't see you. I will take all the responsibility for whatever happens."

A bit later, when the Soviets saw that the training was going smoothly, they offered to help us without pay. They wanted to jump in, perhaps to claim credit. I responded that if I met any obstacles that I could not handle by myself, I would ask for their help. But I didn't need any support just then. We were having 500 men jump per day. The training was completed ahead of the schedule without any difficulties.

At the end of the training program, we had created one airborne brigade, all well-versed in military science, physically and mentally strong, disciplined and self-reliant. This was all done without the help of any foreign government, and without asking for any additional budget except for food, clothing, fuel, parachutes, and stipend.

In March 1983, the unit with its inaugural graduates was named the 5th Airborne Brigade. The day after the graduation, the brigade was given the mission of assaulting the hideout of bandits in Bale province. These bandits had never been challenged until then. The airborne troops rappelled down the cliff, dropped amidst the bandits, attacked and captured several of them, and their weapons and ammunition. The rest were scattered.

As soon as the 5th Airborne Brigade completed the Bale mission, it was sent to Afar. There it was assigned the task of taking direct control of the road between Assab and Addis. The airborne brigade quickly resolved the security problems there. The Assab-Addis road could now be used both day and night. The 5th Airborne Brigade also wiped out the bandits in Wello, enabling the inhabitants of Wello to live without fear. Moving from Wello to Tigray, it also fought several battles against the Woyane and recorded victory in a short time.

The 5th Airborne Brigade, at a small cost to the treasury, and in a very short time, had trained and fielded a thoroughly competent force, which attained immediate and decisive results. Seeing this, the Derg appointed me to create three additional brigades from the people's militia at Tatek. The brigade grew to a division.

A briefing of the high-level team led by the Minister of Defense, 1983, concerning the proposed site for Headquarters of the new Airborne Division.
Author: Front row, 1st left-pointing.

CHAPTER 23
Red Star Campaign

Soon after Ethiopia won the Ethio-Somalia war in the east, the Derg was determined to find a lasting solution to the civil war in Eritrea, so that the people of Eritrea could live without fear and become productive citizens.

The plan to control Nakfa and conclude the war in a short period was not successful because the separatists kept opening new fronts, forcing the Ethiopian military to spread itself too thin and shift from offensive to defensive. This conflict led to the concentration of military power in Eritrea, creating a significant strain on the economy of the country with the massive need for food, clothing, medical care, fuel, ammunition, and salary for the military.

Below is an example of the resources allocated in Eritrea in 1980-81.

Troops	
Regular troops	21,000
Peoples Militia	120,000
Total	141,000

Weapons	
122 & 130 mm Heavy Guns	621
B.M. 21	37
82 mm Mortars	961
Mitrailleuses	5,432
AK-47 Rifles	90,000
T-55 Tanks	159
Armored Cars	153
Heavy and Light Vehicles	3,000

Annual Ammunition Consumption	
130 mm Heavy Gun	12,089
122 mm Heavy Gun	43,059
BM 21	20,739
AK-47 Rifles	5 Combat Loads

Annual food consumption

404,488 quintals (1 quintal = 100 kgs) of food of various types, estimated to cost about 61,874,000 ETB were issued to the troops.

Clothing

Clothing worth about 13,309,34 ETB was issued to the troops.

Fuel, oil, grease

The annual consumption of fuel was worth of 19,434,982 ETB

Gravel road construction

Engineers constructed roads worth about 613,279 ETB.

Transport

In just one year, 500 heavy trucks transported 100,000 troops and 153,900 tons of materials from Addis to Eritrea. The total estimated cost of this was 36,678,159 ETB. The total estimated average cost was 1.2 million ETB per day. The Red Star campaign was proclaimed in 1982 to reduce the continuous daily spending of millions of ETB and the loss of thousands of lives. Therefore, the primary objectives of the Red Star Campaign were to:

- End the Eritrean civil war once and for all and liberate Eritrea from separatists.
- Repair or replace infrastructure damaged by the war, including factories, government offices, schools, health institutions, roads, banks, water, and power lines.
- Restart water, electricity, telephone, and bank services quickly.
- Make government employees productive since the government was still paying their salaries although they were often unable to work during the war.
- Ensure students returned to their classrooms at the earliest time possible.

The Derg instructed all government department offices in Addis to transfer their annual budget, except their salaries, to Eritrea, and to expedite the construction and repair of infrastructure. The government was also committed to providing subsidies since the yearly budget for the campaign was not enough.

The Derg moved all the officials of the different sectors to Asmara to supervise the ongoing work and to ensure its completion in a shorter period. All the efforts by the central government made the Eritrean people happy and gave them some hope for their future. Factories were reopened, and workers began to be productive again. Students started going back to their classrooms. Organizations around Asmara resumed their services and repairs were quickly done.

The work of transporting the fighting men and supplying weapons, ammunition and other provisions were being done by air and land, from Addis to Asmara. The troops that were organized formerly as Task Forces were now organized under divisions, corps, and Armies.

The force assigned in the Algena front was named "Woukaw," which is the Amharic phrase for "hit him hard." The command at Nakfa gate was named "Nadew," the Amharic phrase for "destroy him," and the front was "Mebrek," the Amharic word for "Lightning," and the command around Asmara, "Mekit'" which means "Put up a strong defense."

Before the start of Red Star Campaign, President Mengistu Hailemariam called a meeting to hear a report on the preliminary preparations of the campaign from the commanders of each unit involved.

During the meeting, most of the commanders reported that their troops were well prepared. The commander of the 21st Mountaineer's Division, which had come from Ogaden, stated that his soldiers had made sufficient preparations and that he did not have the slightest doubt that the Red Star Campaign would be successful and that the president should not worry.

Brigadier General Abera Abebe, Commander of the Woukaw command said that the report just given by the 21st Mountaineer's Division commander was not correct. He went on to say that the 21st Mountaineer's Division was less than a year old and had been in

Eritrea for less than one month, did not know the terrains of Eritrea, and had no operational experience in dealing with the separatists. Just the fact that the division is named "Mountaineers," does not mean it can overcome the mountains or perform miracles. The troops needed training. There are troops with no experience in Algena front, too. They should both be given more time for joint exercises together with tank and Artillery. The soldiers are also not well prepared mentally. We should not be in a hurry to launch this campaign.

The President rejected General Abera Abebe's proposal and ordered that the Red Star Campaign be launched immediately, and the following battle order was given to various commands.

Nadew Corps Front Command:

Composed of the 17th Infantry Division, the 18th Mountaineer Division and the 22nd Mountaineer Division reinforced with tanks and Artillery, positioning itself at Nakfa gate, under the leadership of Brigadier General Woubetu Tsegaye, at hill point 1702, would prevent the separatist reinforcement troops from getting to Algena by closing the road from Nakfa to Algena. It would take control of Nakfa by seizing the high grounds to the left and right of Hill 1702 at the Gate of Nakfa.

Mebrek Corps Front Command:

Composed of the 21st Mountaineer Division, the 24th Infantry Division and the 2nd Infantry Division, reinforced with tanks and artillery, and under the leadership of Brigadier General Kumelachew Dejene, would assault Asmat and Terit and take control of Kerkebet.

Mekit Corps Front Command:

Composed of the 14th Infantry Division and various attached of brigades, strengthened with tanks and artillery, would secure Asmara and its surrounding areas.

Woukaw Corps Front Command:

Composed of the 15th Infantry Division, the 19th Mountaineer Division and the 23rd Infantry Division, reinforced with tanks and artillery, and under the command of Brigadier General Abera Abebe, would position itself in the Algena front and assault Katar, Djebel, Dembebit, and Atsebha valley and take them under its control.

The divisions under the Woukaw command were given the following directions:

The 15th Infantry Division would attack and take over Djebel and Dembebit and use the road from Sudan to Algena to prevent the separatists from bringing reinforcement from Sudan to Algena.

The 19th Mountaineer Division, passing South of Katar would advance to Adobaha, then attack and take the Adobaha Valley under control. The 23rd Infantry Division would assault and take Bilkat under its control using the road leading from Nakfa to Algena. It would prevent separatists from sending reinforcement troops to Algena.

The campaign immediately commenced on both fronts.

The mission given to Woukaw commander and the 15th Infantry Division, to assault Djebel and Dembebit and to put them under control, was accomplished. They attacked both hills and quickly took control of them. They then closed the Sudan-Algena road using anti-tank mines and checkpoints.

The 23rd Division was unable to take control of Bilkat. Even though the Ethiopian soldiers could not seize Bilkat Hill, they still stopped the separatist reinforcement troops from entering into Algena and taking Katar and Megareja under their control. The 19th Mountaineer Division could not cross Bilkat and reach Adobaha.

Nadew Corps Front Command:

The Nadew Commander assigned the 3rd Infantry Division the task of taking the high grounds, east of Nakfa gate and closing the gravel road from Nakfa to Algena so that the separatists couldn't use it to reinforce their troops in Algena. The 17th Infantry Division was ordered to take the strategic high grounds in front of Nakfa gate under its control, and thus create a suitable condition for the 18th Mountaineer Division to launch an assault on Nakfa. The 15th Infantry Division, also taking advantage of the opportunity created by the 17th Division's attack, would seize control of the town of Nakfa. The 22nd Infantry Division was assigned to guard Afabet and feint an attack on Halhal.

The 17th Infantry Division launched its assault on Hill 1702 and succeeded in temporarily taking control of it. But, was unable to turn back the counter-attack of the enemy and was forced to retreat. In a second attempt, the 3rd Infantry Division regrouped with the 18th Mountaineer Division and attacked Hill 1702. Although they assaulted Hill 1702 repeatedly, they were not successful.

In the third phase of regrouping, all the divisions launched attacks on their respective fronts and gained some ground but produced no definitive results.

In the fourth phase, after regrouping, the 2nd Infantry Division, the 21st, and 22nd Mountaineer Divisions, and the 24th Infantry Division launched an attack along their respective fronts in an attempt to take the militarily advantageous high ground but had no success.

The Mebrek Corps Command:

The Mebrek commander gave the following tasks to each division:

- The 21st Mountaineer Division was ordered to attack Koor and put it under control.
- The 24th Infantry Division was ordered to attack Asmat and put it under control.
- The 2nd Infantry Division was ordered to attack Halhal and put it under control.

The 21st Mountaineer Division attacked and held Koor, but it was not able to stop the enemy's counter-attack and so lost Koor and retreated. The 2nd Infantry Division briefly controlled Halhal, but it abandoned Halhal and withdrew when it saw that the 21st Mountaineer Division was retreating.

Brigadier General Kumelachew, Commander of the Mebrek and

Colonel Woubeshet Mamo, Commander of the 21st Mountaineer Division were held responsible for the failure of the Red Star Campaign and for not carrying out their responsibilities. An investigation team from Asmara arrived in Kerkebet. Following the report from the investigation team, both commanders were taken to Asmara by helicopter to appear before the military court. Brigadier General Kumelachew was sentenced to imprisonment, and Colonel Woubeshet Mamo was sentenced to death. A firing squad later executed Colonel Woubeshet in front of his troops.

At this point, the Derg decided to terminate the Red Star Campaign and begin planning for another campaign.

While the Red Star Campaign was launched and being fought in the north, I was in Addis handling the challenges of establishing an airborne division with the militia who did not meet the academic requirements necessary to be trained as airborne men. When I was appointed as the commander of the new airborne division and airborne training school, I requested that I be allowed to recruit men based on the criteria I had prepared. However, I was told to select only men from the militia at Tatek. Most of the militia met the physical requirement, but many of them did not meet the academic standards.

Accepting the order, I took a team from the airborne training school and left for Tatek. After we completed our selection, trainers got the men started with basic training at Tatek so that the recruits weren't idle. I went to Awassa with another team to check and see if the old training ground on the shores of Lake Awassa, were still available for our use.

When I returned from Awassa to Addis, the Chief of Staff, Major General Hailegiorghis Habtemariam, called me to his office. There I was informed that the government had decided that the recruits should be trained in Eritrea. I was to fly to Asmara the next day to meet with Major General Merid Negussie, Commander of the 2nd Revolutionary Army, and work out the details of training and return to Addis as soon as possible.

Major General Merid Negussie briefed me that Asmara was in a seriously dangerous situation because they had no reserves. The airborne division could not all train at the same time. While one brigade was training, the other two brigades would have to defend Asmara and its surrounding areas. One of the two brigades would man the trenches; the other would be held in reserve. The units would rotate through each assignment as the training progressed.

I told him that the recruits had not yet received any training and that the Chief of Staff had told me to bring them here for training, not for deployment. This instruction had come straight from the President.

Major General Merid told me that he had since spoken with the Chief of Staff and that this was the latest decision. Since I could not argue with him any longer, I followed his orders. We selected Dekemehari to train the first brigade, and I gave the list of materials and supplies that I needed to the relevant departments.

As soon as I returned to Addis, we began transporting the troops to Asmara. The first brigade entered the trenches as soon as it arrived. The 2nd Brigade was assigned as a reserve as it underwent training. The 3rd Brigade continued infantry training in Dekemehari.

I tried to remind Major General Merid that the best use of airborne troops was not defense.

"If Eritrea is gone, what good is the airborne unit?" he stated. "Eritrea is in a precarious situation. We should be discussing how to save Eritrea not what the airborne troops should do or should not do. We, the commanders, should also be ready to go into the trenches and fight."

CHAPTER 24
The Areza Operation

Although the earlier Red Star Campaign had not been successful, the Ethiopian forces were still ordered to reinforce their defense zones and prepare counter-attack to take control of the high grounds. In a surprise attack, the Ethiopian forces managed to take Melebso, Asmat, Koor, Terit and other key areas under their control. Unfortunately, shortly thereafter, the separatists launched a counter-attack of their own and took these areas back.

From 1982 to the beginning of 1984, both sides kept taking turns controlling these areas.

The separatists used the Tesenei-Zulei road to receive logistical supplies from the Middle East and their allies abroad. The Tesenei-Zulei road passes through Barentu, Bosheka, Mulki, Badme, Areza and Mendefra to get to Zulei port by the Red Sea in Akale-Guzay province. The separatists used the road to recruit teens from the highlands and send them to their training center in Sawaforto. This same road was used by the separatists to send infiltrators to agitate the peasants, the youth, the workers and the people. They would tell them not to let themselves be organized into farmer associations and not to send their children to the national military service. Those who didn't listen could face kidnapping or assassination.

By placing a brigade at Mulki, the Ethiopian forces prevented the separatists from having access to the vital Tesenei-Zulei road, disrupting their recruitment and harassment, and cutting off their supply line. The separatists were denied free movement, contact with the population, and their supplies from Zulei. In response, they launched a surprise night attack and dislodged the Ethiopian forces from Mulki after a brief battle. The Ethiopian forces withdrew back to Areza which is closer to Asmara, and from where they would have easier access to supplies and reinforcements if attacked again.

Brigadier General Samson Haile was assigned as coordinator of the 14th Division and the 2nd Infantry Divisions. Areza is fifteen kilometers west of Mendefera and well known for its barley and wheat production. Mendefera is in Seraye Province, fifty kilometers south of Asmara. The two infantry divisions led by General Samson stopped separatist movements in the area, and the Tesenei-Zulei road was again under control. The road between Tesenei to Zulei road was a matter of life and death for the separatists and critical for them to continue their struggle. They could not continue their fight without weapons and ammunition. Therefore, the separatists and the Woyane agreed to form a united front and launch an attack with a combined larger force to retake the road at Areza.

On November 24th, 1984 at 0100 hours, three separatists and Woyane brigades closed the Mendefera-Areza road, then launched an attack on Areza. The two divisions fought back the whole night of November 25th, 1984 but could not dislodge the separatists. on Brigadier General Samson's request, Major General Merid ordered Brigadier General Regassa to take the 52nd Para-Commando Brigade and the 7th Airborne Brigade as reinforcements to the battle in Mendefera. I was also ordered to assist Brigadier General Regassa.

The 7th Brigade and the 102nd Airborne Division were both recruited from the people's militia at Tatek and brought to Asmara in September 1984. The troops were peasants with airborne uniforms and red berets and had no training. They were told that from then on, they were airborne troops. They were assigned as reserves of Mekit Corps. However, we quickly introduced them to individual field crafts, squad-tactics, and basic weapon training.

The 5th Brigade was the only airborne brigade in the whole of the 102nd Airborne Division with proper training and some fighting experience. I had repeatedly asked that we bring the 5th Brigade from Wello to its parent unit, to help us train the other brigades. My request was ignored. The politicians and commanders claimed that the 5th Brigade was too important for the existence of Wello. On November 25th, 1984 at 1500 hours, the 52nd Para-Commando and the 7th Airborne Brigade moved from Asmara to Areza via Mandefera. The two brigades arrived at the outskirts of Mendefera at a town called Adigorento.

After launching a flank attack against the delaying force sent by the separatist that we encountered on the way and forcing them to disperse, we arrived in Areza on November 25th, 1984 at roughly 2300 hours. Brigadier General Samson and his troops were in an intense battle when we arrived in Areza. He told us that the separatists were approaching, and he urged us to get to him quickly before the separatists demolished his command post.

We assured him that we would do all we could to get to him quickly, but that the terrain was rugged. Moving in pitch black night by itself is dangerous but becomes much riskier when the ground is covered with slippery rocks, fragmented stones, and thorny bushes. We had to make sure we made no noise. Walking on the balls of our feet, we tried to avoid stepping on dry leaves or tins. We removed everything that might reflect light, such as wedding rings or watches. We had to move slowly so that no troops got separated and left behind.

We waited until daybreak, so we could distinguish the territory held by the enemy from that held by our own forces. I gave orders to the 52nd Para-Commando Brigade to attack from the east, and I went with the 7th Brigade to attack from the north. Since we could advance with only one battalion at a time, I gave the order that the two brigades were to launch an attack along their front with just one battalion. As we moved forward, we could then deploy the second battalion. I gave instructions to the 52nd Para-Commando brigade to open fire at a previously designated time to distract the separatist's attention away from the north. They opened fire at 0545 hours. The separatists' attention was concentrated to the east, and they did not expect an attack from the north, thinking the terrain was too unsuited for launching an attack. At 0550 hours, our attack from the north encircled the separatist, forcing them to retreat west.

The two brigades chased the separatists towards the west. General Regassa moved his command post from Mendefera to Areza, stationed the 2nd Division in Areza, and ordered to the 14th Division to move to Maidema and join the two brigades in the pursuit. The 8th Airborne Brigade moved out from the Asmara trench and assaulted the western ridge of Maidema and took that enemy stronghold under control. The separatist began to defend from the Hill 1499, known as Tsadi-Adi Hill.

On November 25th, 1984, the 16th Sentik Mechanized Brigade was ordered to move from Barentu eastward to Mulki to prevent them from retreating to the western desert.

The Sentik brigade was composed of:

- One tank battalion with 31 tanks.
- One 122mm artillery battalion, with 18 guns.
- One armored vehicle battalion with BTR's and BRDM's.
- One anti-aircraft battalion with ZU-23 anti-aircraft guns.
- Several tankers for fuel and water.

The separatist gathered its forces from Algena and Nakfa fronts and moved towards Bosheka and encircled the Sentik mechanized brigade. It would be a catastrophic loss if all the weapons of Sentik Brigade were captured by the separatists. On November 25th, 1984, the 5th Airborne Brigade started traveling at night, from Korem in Wello to Mekele by land, and by air from Mekele to Asmara. Its mission was to prevent the Sentik Brigades from being defeated and captured.

I left Areza to receive Lt. Colonel Girum Abebe, the 5th Airborne Brigade's commander, at the airport. He had arrived on the last flight on November 27th, 1984 at 1530 hours. From the airport, we went to the 5th Airborne Brigade base. Soon after, Maj.General Merid Negussie joined us there while the troops were having their meal.

Colonel Girum and I approached and saluted General Merid while he was still in his car. He got out and said to Colonel Girum, "Welcome, how was your journey?" After a few steps, he stopped and addressed us both. "What should we do?"

"What's happened, sir?" I asked.

"The 16th Sentik Mechanized Brigade has been encircled by the separatists. If it is eliminated, it means the end of the Eritrean war."

"What do you want us to do?"

"I know that the 5th Airborne Brigade has had a long day and is exhausted, but it is my wish that they spend the night ten kilometers from Asmara.

Colonel Girum and I looked at each other. He understood that I wanted him to respond. Colonel Girum then turned to General Merid.

"If we are provided with vehicles, we can move right now, there is no problem on our part."

Moments later, before the troops were done with their meals, the trucks started arriving.

"Those of you who have finished eating, get armed, fill up your water canteens, and get on the trucks," Colonel Girum ordered.

Then we realized that General Merid, although he had asked for our views earlier, had already decided that the 5th Airborne Brigade would move that same night, and had already made the transportation arrangement.

In less than ten minutes, the troops were armed and, on the trucks, and had started their journey towards Keren at 1630 hours. Some of the troops did not finish eating and some were still waiting in line to get their meal when they stopped to get armed.

The 5th Airborne Brigade spent the night not ten kilometers from Asmara as previously said, but rather travelled ninety kilometers, and spent the night at Keren. On November 28th, 1984 at 0400 hours, it left Keren, passed Akordat and Barentu, and reached Beshoka at 1400 hours.

The Sentik Mechanized Brigade had been fighting for two consecutive days with no resupply or reinforcement. The men were showing signs of despair because they did not know if reinforcements were coming to their rescue. The Sentik Brigade was the reserve unit itself, so who was to rescue the rescuers? On the other hand, the separatists had been fighting in shifts, and were able to replace and resupply their troops.

The 5th Brigade arrived at Beshoka at practically the last minute, just as all seemed lost. The separatists were about to launch their final assault on the Sentik Brigade's dispirited and exhausted men, who they had closely encircled and were within grenade-throwing range. The separatists were about to make their charge. The Sentik Brigade was finished.

The 5th Airborne Brigade troops arrived in Beshoka and quickly jumped down, lined up, and went into battle. The 5th Brigade does not hold ground and fight. They attack. They go for direct contact; grenades, bayonets and hand-to-hand combat. The 5th Brigade's attack was completely unexpected by the separatists, forcing a hole in their circle. The 16th Sentik Brigade and the 5th Airborne joined up and rolled back the encirclement, creating a corridor on one side.

The fight switched from bullets to grenades. The 5th Airborne fighters caught the enemy's hand grenades in the air like so many balls and threw them right back at them. Not one grenade exploded on our men's side. The separatists were quite alarmed. They couldn't, as our men had done, catch and throw back the same grenades because a hand grenade has only three seconds to explode once the pin is removed. Still, the separatists continued to combat the airborne brigade. The fight went on from November 28th, 1984 at 1400 hours until November 29th, 1984 at 0300 hours. The tanks, armored cars, heavy guns and vehicles of the 16th Sentik Mechanized Brigade were withdrawn from the previously encircled position and moved back to give fire support.

On November 29th, 1984 at 0320 hours, the separatists threw themselves into a final, all-out counter offensive.

Lt. Colonel Girum, noticing that some of his troops were faltering, shouted to those trying to retreat, "Where are you going? Airborne soldiers do not retreat! Move forward! Follow me!"

He fixed his bayonet, removed the safety pin and threw a hand grenade, then moved towards the enemy. His troops followed him into battle. The enemy had to retreat. During the thick of battle, Lt. Colonel Girum was everywhere, encouraging his troops, giving orders, and helping the wounded and the dying.

After few minutes, Lt. Colonel Girum went to his knees and then laid flat on the ground. As his troops tried to lift him up, he told them in a weak voice,

"I'm all right, you go continue the fight."

His troops first thought that he had fallen from fatigue because he had not had any food or water for forty-eight hours. But when they tried to move him, they found that Lt. Colonel Girum's chest was soaked with blood. That is when the troops realized that their commander had been hit by a hand grenade fragment during fighting. He had not even been aware of it. His troops, worried, assembled around him. Again, he told them,

"I am all right. Don't worry about me. Just set me down here, then go and finish the fight."

Some of them remained behind to watch over him, while the rest of them moved toward the fighting.

His men were back in less than ten minutes. Lt. Colonel Girum was dead without seeing the result of his valiant struggle. The troops of the 5th brigade wept bitterly at the loss of their beloved and courageous commander. They had looked up to him as a father and friend. They wished their commander could have seen their victory, finally. The sacrifice of Lt. Colonel Girum Abebe was not in vain. The separatists were defeated, and Eritrea was not lost. The armed forces were saved from a humiliating defeat.

As soon as I heard of the death of Lt. Colonel Girum Abebe, I flew by helicopter from Mulki to Bosheka. I spoke to calm the troops and appointed Major Kassay Tadesse, who was the most senior officer, to take over the command. The troops asked if their commander's body could be sent to his family in Addis. But, it was government policy that everyone killed on the battlefield, regardless of rank, was to be buried on the battlefield. There were too many men killed in action every day and transporting all the bodies back to Addis would be too costly. Any special arrangement for the bodies of officers or commanders would create dissatisfaction among the lower ranks.

The battle of Bosheka and Mulki ended with our triumph. The separatists retreated. The 14th Infantry Division took over Mulki and the 2nd Division took over Areza. The 5th Airborne Brigade was flown from Barentu to Asmara on December 8th, 1984. Between December 26th and December 27th, 1984, the 52nd Para-Commando Brigade was ordered to march to Asmara.

The 7th Brigade cleared the way from Areza to Mainefi. On the way, they found a large quantity of separatists' property. They took with them as much as they could carry to Asmara and burned the rest. The 6th and 8th Airborne Brigades, after the battles of Maideme and Mulki, returned to Asmara by road.

Below is 5th Brigade's casualties in the battle of Bosheka. I have no figures regarding casualties on the part of the 16th Sentik Brigade.

Officers Killed	3
Privates	70
Total	73
Officers Wounded	7
Privates	193
Total	200
Killed	217
Wounded	380

Source: my own notes and information from the operations and intelligence department.

The battles of Areza, Mulki and Besheka were complete victories for the Ethiopian forces. The day after the battle ended, I returned to Asmara and resumed training the 102nd Airborne Division.

I received the following letter of appreciation from the Commander of the 2nd Revolutionary Army, Major General Merid Negussie.

No ሁኢሠ1/6/1111/

(Top Secret) 2609/4168. *Dec. 27th/1985*

To: Col. Tesfaye HabteMariam

The campaign we carried out to dislodge and eliminate the Eritrean bandits from the strategic grounds they held in Seraye Province at Areza and Mulki was successful and we came out victorious. However, the bandits fought extremely hard, eventually compelling our forces to take a defensive position at Areza and stay there.

Once again, the bandits gathered additional force and attacked to eliminate two divisions of the Ethiopian forces at Areza, and to capture a large quantity of equipment. They then marched to Mendefera, and arrogantly launched an encirclement assault at Abrit, Areza on November 25th, 1984. They closed the line of retreat of the Ethiopian forces and their supply line, thus encircling the Ethiopian forces at Areza. The situation was going to result in regrettable failure of our country and our Ethiopian military.

The 7th Airborne Brigade under your command was among the troops sent to rescue our forces encircled at Areza and eliminate the bandits. You arrived at the battlefield, and with your usual determination and courage you coordinated the battle jointly with Brigadier General Regassa Djima. You, together with other comrades, formulated a strategy, directed our soldiers wisely and heroically, and got the Mendefera to Areza road opened. Your actions safeguarded the two infantry divisions and their equipment that were encircled at Areza, thus contributing highly to the victory obtained. It is with great pleasure that I take this opportunity to express my utmost gratitude to you.

You risked your life for your country, entered an intensive battle, provided determined leadership and proved yourself one of the heroes of the victory. I once again express my admiration to you and express my wish that all your future struggles will be successful.

Merid Negussie, Maj. General

There are four battalions in an airborne brigade. I finally got permission for my repeated request to take one Brigade at a time to Debre Zeyit for airborne training. The airborne training school at Debre Zeyit only had the resources to accommodate one battalion at a time. I had promised to complete the six-month training in no more than two months and return them to Asmara.

Two of the four battalions, while waiting their turn to train at Debre Zeyit, would live in make-shift tents near Koka Dam, some twenty kilometers from Debre Zeyit, and finish up their infantry training.

The other two battalions would be at Debre Zeyit. One battalion would be doing jump training while the other would learn how to pack their parachutes, practice proper parachute landing falls, and complete the physical commando training in the one of the several areas set up for this purpose around Debre Zeyit. This would shorten the time it took for them to go through jump school.

After completing their jump training, they would be taken to Koka to continue their infantry training, while the other two battalions that were at Koka would now come to Debre Zeyit training school to conduct the training that their comrades had just completed.

The four battalions, working in shifts for two months, completed the six jumps required to qualify them as Airborne soldiers. They did five of the jumps during day time and one of them at the night. After completing their training, they returned to Asmara proudly wearing their red beret and the silver wing on their chests.

We continued the training program in a similar manner by bringing the two other brigades from Asmara to Debre Zeyit in shifts. The training of the three brigades in shifts, which started on January 23rd, 1985, was completed on August 6th, 1985. The morale of the troops was high after they completed their jump training. Their confidence and fighting spirit grew and they became more self-disciplined.

While the 102nd airborne division was still under training in Debre Zeyit, the separatist dislodged the Woukaw command from Algena by pooling all their resources and attacking using mortars, rocket launchers, and machine guns. They then placed, under their control, the towns and agricultural estates found in Gash and Setit provinces.

On July 6th, 1985, they put Barentu under their control. The Ethiopian military forces now came up with the next operation to curb the swift attacks of the separatists.

They called it Bahre Negash.

CHAPTER 25
Bahre Negash Operation

THE PRIMARY OBJECTIVE of the Bahre Negash Operation was to liberate Barentu from the separatists. For this, it was necessary to control the high ground in Bosheka and Mulki on the route to Barentu. The 14th Infantry Division was ordered to move from Areza to Mulki, and the 102nd Airborne Division was ordered to take over Areza from the 14th Division. A few days after it took over Areza, the 102nd Airborne Division was ordered to clear both sides of the area extending from Areza to Mulki, and to enter Bosheka via Mulki. The district extending from Areza to Mulki is surrounded by a chain of mountains. We spent the day going up and down mountain roads until we reached Mulki at night fall.

Mulki is on low ground, surrounded by hills. One week had passed since the 14th Division had entered Mulki. Although it was dark, we could see a series of prominent heights at a short distance from our military camp, but it was too dark to move on and acquire the high strategic ground. For security purposes, we decided to spend the night where the 14th Division had set up camp. I spent the night close to one of the brigades after informing the commander of the 14th Division, Lt. Colonel Getahun Wolde Giorghis that we had arrived.

Overnight, the separatists dispatched a force, which got into position at Mulki and prepared to attack early in the morning. They did not know that the 102nd Airborne Division had arrived at Mulki and was spending the night there. Early in the morning, while we were preparing to continue our journey to Bosheka, they started firing shots from the nearby hills.

I ordered the brigade commander next to me to take the hill to the east under his control and commanded the other brigade commanders to take control of the other hills to the north and west. Platoon and company commanders rushed towards the hills with their troops. I climbed to a higher position so I could see the battle and give directions.

The airborne troops, after a brief battle, managed to seize the hills to the east and south. But, we still had to fight for some time before we took the high ground to the north and west since that was the direction from which the separatist attack was coming.

The separatists retreated to Bosheka and continued their journey towards Barentu. The 102nd Airborne Division entered Bosheka and reorganized. The 14th Infantry Division left Mulki and entered Bosheka, following the 102nd Division and established provisional control. For the second time, the 102nd Airborne Division made great sacrifices to save the 14th Infantry Division from annihilation.

The order in which our forces lined up for the counter-attack, to retake Barentu, was as follows:

- The 102nd Airborne Division.
- The 3rd Infantry Division.
- The 14th Infantry Division.
- The 17th Mechanized Brigade
- One tank battalion.
- One artillery battalion.

The 3rd Division held a position at the base of a mountain located east of Barentu as its assembly area. The 14th Infantry Division held the area starting from the Barka River to the boundary adjacent to the holding of the 102nd Airborne Division. The 102nd Airborne Division, together with the 17th Mechanized Brigade, positioned itself with the main attacking force in front of Barentu town. The tank and artillery gun battalions took up their firing positions behind the 102nd Airborne Division.

On August 22nd, 1985 at 1830 hours, the 3rd Infantry Division would start out from its assembly area and take the mountain east of Barentu under its control by August 23rd, 1985 before 0500 hours.

On the same day, the 14th Infantry Division, at 1830 hours was positioned to defend the area extending from the Barka river to the left of the area held by the 102nd Division. On August 23rd, 1985 at 0530 hours, the Airborne Division, together with the 17th Mechanized Brigade prepared to launch a frontal attack and take Barentu under its control.

Following to this plan, the 3rd Infantry Division commander started out from the assembly area and moved towards the mountain. After one hour, he was to report his location by means of prominent objects or map grid reference. At 2000 hours, halfway up the mountain, he radioed his position.

At 2130 hours, shots were heard at the base of the mountain in the assembly area. When the commander of the division was asked what the shot was, he replied that the enemy was about to cut him off from behind and that reinforcements should be sent to him. The 8th Airborne Brigade was sent to his position by the 3rd division commander.

The 8th Airborne Brigade raced to reach the 3rd Division's position, about ten kilometers away. He got there in less than two hours. The 8th Airborne found not the 3rd Division, but a separatist ambush. They waited until the 8th was in the kill zone, then brought a deadly hail of fire from mortars, rocket launchers and heavy machine guns. The destruction was unbelievable. There was blood everywhere and many men were lost.

The brigade commander himself died bravely with his troops during the fighting, taking with them as many as they could. Those who escaped death fell into the hands of the enemy. Reinforcement troops were sent, and the battle continued throughout the night. On August 23rd, 1985, at 0530 hours, the 3rd Army Division was ordered to counter-attack and take the mountain under their control, which they did.

The separatists brought all their forces to defend Barentu town. On August 24th, 1985, the 102nd Airborne, together with the 17th Mechanized Brigade, launched a frontal attack and eliminated the separatists, and took Barentu under control.

The Ethiopian forces continued chasing the separatists. On August 25th, 1985, they captured Sawaforto, the separatists training center, repair workshop, and depot. They captured spare parts for tanks and heavy trucks, electric power generators, agricultural equipment, medical equipment, medication, fuel, and much more.

The fighting continued until August 30th, 1985, when the Ethiopian forces controlled almost all the towns and real estate found in the Gash and Setit provinces, including Haikota, Tesenei, Aligider, Sebderat and Galodge. The separatists lost much of their strength and equipment. Their military was demoralized, their fighting spirit was lost, and they scattered in several directions. Those who were on the ground later stated that the Ethiopian military could have taken Nakfa under control without any losses but elated by our victory we celebrated with dancing and music. This gave the enemy a chance to regroup its scattered troops and to re-establish itself.

The first part of Bahre Negash was concluded with a victory for the Ethiopian forces. I entered Asmara together with the 102nd Airborne for the preparation of the second part of the operation.

A new task force known as "Bergid" (the Amharic word for "break open") was established under the command of Brig. General Woubetu Tsegaye to make the second part of the Bahre Negash operation successful. It was composed of the 21st Mountaineer Division, the 23rd Infantry Division, the 102nd Airborne division, one tank battalion and one heavy gun battalion.

To help the Bergid task force carry out its mission successfully, the 102nd was given the following task:

- One airborne brigade was to be lifted by helicopter from Afabet to Algena and land on Katar mountain.

- One airborne brigade was to be lifted by plane from Asmara to Algena and jump by parachute.
- One airborne brigade was to travel by landing craft from Massawa and enter Algena.

After receiving their orders, the 102nd airborne division started making its preparations. One Brigade would be sent to Afabet for helicopter training and complete all the logistics and administrative work by September 4th, 1985, at 1600 hours. As a M-8 helicopter could accommodate only thirty soldiers at one time, therefore each team was organized with thirty people or less. Ten M-8 helicopters could lift three hundred people at one time. The ten helicopters would have to make four flights each, a total of forty flights to transport 1,200 people. The flight from Afabet to Algena took thirty minutes one way or one hour round trip.

If the first flight left Afabet at 0530, the men would be on the ground at 0600. The second flight would then arrive in Algena at 0700, the third flight would arrive at 0800 hours, and the final flight would touch down at 0900 hours.

Ten teams of thirty soldiers would be made ready for each helicopter. Each team would be given a flight number and a helicopter number, and the jump masters would ensure that all activities were carried out efficiently and quietly.

The brigade in Asmara would prepare for jumping. Parachutes would be brought to Asmara from Debre Zeyit. An Antonov 12 airplane, with the capacity to carry three hundred parachutes on each flight would require five flights to transport the 1,500 parachutes from Debre Zeyit to Asmara. An Antonov could make two flights a day from Debre Zeyit to Asmara. We requested three Antonovs from the Air Force and the 1,500 parachutes were transported Debre Zeyit to Asmara in one day.

The parachutes brought from Debre Zeyit had been in storage and had gone unused for more than three months. According to the rules and regulations of parachute use, the packed canopy of the parachutes will have to be taken out of their canvas, stretched out, aired and inspected by trained parachute rigger. The jumper would pack his own parachute under the supervision of a parachute rigger to make it ready for use.

One plane can take seventy jumpers at one time. The dropping area could accommodate thirty-five jumpers at one time. Two teams are set up so that the plane may drop the seventy jumpers in two rounds. There were 1,200 jumpers prepared for this mission. Six Antonov 12 airplanes, each making three flights, a total of eighteen flights were required.

This was the drill:

- Boarding time of jumpers at Asmara airport, 10 minutes.
- Flight time from Asmara to Algena, 40 minutes.
- Making two teams jump on to dropping area, 5 minutes.
- Return time from Algena to Asmara, 40 minutes.

The round trip from Asmara to Algena, for one round of jumping would require 95 minutes or one hour and 35 minutes for three flights. A total of four hours and forty-five minutes was required to get the job done.

If the take-off time for the airplane from Asmara airport for the first round is 0530 hours, the finishing time for the last round jump would be 1015 hours. All the jumpers would rig up their parachutes and place them in the store under the supervision of the riggers. The jump masters would record the names of the jumpers and the serial number of the parachutes on the jumping form.

Four jump masters were assigned for each airplane. All organizational and administrative tasks would be completed the day before the jump by 1800 hours.

To expedite the re-grouping in the assembly area each platoon was structured independently. Ground exercises and the practice of rigging parachutes was to be carried out south of the airport at Adiguaded on the outskirts of Asmara.

One airborne brigade would arrive at Massawa on October 9th, 1985. The Navy would provide a landing craft for transporting 1,500 men from Massawa to Melkafatuma, a natural port found near Algena.

The arrival date at Melkafatuma was to be D-Day at 0600 hours. Start time from Massawa was decided based on the time required to get from Melkafatuma to Massawa.

Assuming that D-Day was October 10th, 1985, the 21st Mountaineers and the 23rd Infantry Division, together with one tank battalion and one heavy gun battalion would start from Afabet and arrive at Algena on October 10th at 0600 hours.

The first group of 150 airborne soldiers would be dropped at Katar mountain and another 150 soldiers at Tamru mountain at Afabet on October 10th, 1985 at 0600 hours. The transportation and dropping of the soldiers by helicopter would continue. The ones dropped earlier would take the high ground under control and give protective cover to the men dropping next. With reinforcement, they extended their holdings and strengthened them further. The plan was to hold on, even if the enemy put pressure on them, and wait until the 21st Mountaineer's Division and the 23rd Infantry Division joined them. As soon as they arrived at Algena, they climbed the Tamru and Katar mountains. The jumping of one airborne brigade was completed on October 10th, at 1015 hours. The Airborne Brigade that arrived at Melkafatuma on October 10th at 0600 hours and came from Massawa by means of a landing craft, were to reinforce the troops that landed by helicopter on Katar and Tamru mountains.

The 102nd Airborne Division would stay at Algena as a reserve force, while the 21st Mountaineer's Division and the 23rd Infantry Division would march into the Adobaha valley. The Nadew and Mentir command would attack along their assigned fronts and carry out their task on the same schedule.

The troops from Asmara parachuting into Algena would be under my command as 102nd Airborne Division Commander. Lt. Col. Behailu Kinde, Deputy Commander of the 102nd Airborne Division, would take command of the troops coming in by helicopter from Afabet and being dropped at Katar and Tamru mountains. The troops traveling by boat from Massawa and entering Melkafatuma and Algena were to receive orders from their brigade commanders.

The communication with the Division Commander was handled

by encrypted long distance radio at the top of every hour. The brigade traveling by landing craft from Massawa to Melkafatuma would rely on naval communications during its travel. When it reached Melkafatuma, it communicated with the Division Commander by means of encrypted short distance radio. Other brigades, too, when they reached Algena, carried out their communications by short distance (PRC-77) radios. The plan of execution was aimed at encircling and eliminating the enemy. However, the plan had many difficulties from the beginning. A few of the problems encountered during the execution of the planned second mission of the Bahre Negash Operation were as follows:

For unknown reasons, the troops that were supposed to travel by helicopter from Afabet to insert at Algena (Katar mountain) were made to use trucks to get to Algena together with the infantry troops because the helicopters were not provided.

The airborne brigade, that was scheduled to board a landing craft at Massawa and cross the Red Sea and arrive at Melkfatuma on October 10th, 1985 at 0600 hours, reached Melkafatuma at 1300 hours and then arrived at the assembly area at 1700 hours.

We had requested six aircraft to complete the jumping quickly. However, we were provided with only three. The plane that should have carried sixty jumpers now needed to carry one hundred jumpers and we concluded the drops at 1100 hours.

The troops that traveled by vehicle from Afabet arrived at Algena on October 10th, 1985 at 0700 hours, and while they should have rushed to take Katar and Tamru mountains under their control, they stayed at the base of the mountain instead.

The troops that had parachuted down were about fifteen kilometers away from their target. Once again, while it was clear that Katar and Tamru mountains were critical for the takeover of Algena and Adobaha valley, Bergid's commander failed to make use of the airborne soldiers that had parachuted in to control the strategic high ground. Also, he did not provide us with transportation to get us to the assembly area. For this reason, we reached our assembly area at 1600 hours with blood blisters on our feet and severely dehydrated after a strenuous walk on hot sand. Some had to crawl on their hands and knees. Later, when we tried to climb Katar mountain, we found that the separatists had gotten there ahead of us and were entrenched in the Katar and Tamru mountains. They moved into their former trenches and started firing their machine guns on us, preventing us from getting close. It would have been easy to take Katar and Tamru under our control before the separatists got there, but we gave it away because of a failure in leadership.

Our main problem was failing to adapt quickly and come up with alternate plans when our plans, which looked perfect on paper, suddenly proved unworkable on the ground.

The second chapter of Operation Bahre Negash ended in this way with no major victory, except that we now controlled Algena.

On October 25th, 1985, I got an order to return to Nakfa gate for the third mission of the operation.

Seeing that our assaults on the three fronts of the second mission of Operation Bahre Negash had been unsuccessful, our superiors made the following change of plans:

- The Mebrek command at Melebso front and the Bergid command at Algena were to remain where they were and set up an active defense.
- The Nadew command at Nakfa, being reinforced with men and equipment were to continue their assault.
- The 102nd Airborne Division was to move from Algena to reinforce the Nadew command.
- These preparations were to be completed quickly.

Following my new orders, I arrived at Nakfa with my division, the 102nd Airborne. The 22nd and 3rd Mountaineer Divisions were already at their assembly areas.

The preliminary preparation was not yet over. The 102nd Airborne was camped at the rear of the Ethiopian forces, out of range of enemy artillery to study the environment. The area held by the Ethiopian forces was in a valley surrounded by mountains. The gravel road heading from Afabet to Nakfa and the dry, sandy river bed split the valley in two from south to west.

The fact that we were positioned out of the range of the heavy guns gave us a chance to study the topography of the surrounding area and to plan and carry out exercises to deal with the tactics of the enemy, and to adjust to the surrounding terrain as our preparations were being made.

All the high ground in the surrounding areas was under the control of the enemy. The enemy had prepared strong defensive positions, mounted heavy machine guns, and buried mines and obstacles. We had to wait until dark to launch any attack on the mountains where the enemy was entrenched. Any force that carries out a night attack on a mountain with trenches and obstacles must be physically strong, disciplined, well-trained in military science, and have good leadership.

Our training was constant. It continued even on the battlefield when some of our troops went into action. The training program on the battlefield was prepared to deal effectively with the local weather conditions, terrain, and tactics of the enemy.

The 102nd Airborne Division continued its pre-battle practice day and night to carry out the challenging tasks successfully. There was constant assessment done to measure leadership, the execution of exercises, physical fitness and the exercise of discipline. We worked hard to identify our weaknesses and changed them into strengths.

Finally, the preparation for the third mission of Operation Bahre Negash was completed.

Major General Merid Negussie, commander of the 2nd Revolutionary Army, established his command post at Nakfa to direct the operation.

On November 12th, 1985, we were called to the command post to receive operational orders. Major General Merid Negussie briefed us in detail about the current movement and condition of the enemy and the ordered the operations officer to give us our orders.

The first phase of the Bahre Negash Operation Order of the 2nd Revolution Army was to take Barentu under control by a counter-attack. The battle would then continue towards the western desert and liberate the large state farms and the towns in western Eritrean that had fallen under the control of the enemy. It broke the back bone of the separatists by taking Sawaforto, the separatists' training center. We could have taken Nakfa under our control while the separatists were in disarray, but we gave them a chance to regroup while we were rejoicing and snoozing.

The second round of Bahre Negash Operation was a plan for land, air and sea on the three fronts (Melebso, Nakfa, and Algena). We carried out an offensive battle immediately afterwards, but it did not get us the expected results because of problems of execution once again.

We discovered that the main cause of our failure was assaulting in three different directions at the same time. Learning from our mistakes, we made the decision to pool all our forces and assault along one front.

In accordance with this decision, we set up an active defense in Melebso and Algena and carried out an intensive assault at the Nakfa front.

This third phase of Bahre Negash was to eliminate the enemy in the trenches of Nakfa's chain of mountains and to take Nakfa under our control. Our attack would take place at night because the enemy held the high ground and their trenches were heavily protected with heavy machine guns and mined with explosives

For the execution, the 22nd Mountaineer Division was to hold onto their defense position. The first mission of the 18th Mountaineer Division would be to eliminate the enemy camped on the hills found on the right side of the road leading from Afabet to Nakfa, and to take control of Hill 1702 and the high ground in the surrounding area. It was to quickly establish their position and then set up a perimeter defense to stop any counter-attack by the enemy. Then prepared for the next task and await the orders of our superiors.

The 102nd Airborne Division would send one brigade to annihilate the enemy lying low on the strategic ground on the right and left sides of the road from Afabet to Nakfa, then climb to the top of the mountain chain and take control of the high ground in that immediate area. The brigade would establish itself quickly, put up a perimeter defense to stop any possible enemy's counter-attack and wait for our next orders.

Another brigade was to eliminate the enemy lying low at the base of a hill known as "Dalicha Hill." Then climb to the top of the chain of mountains and seize the high ground to help block any counter-attacks by the enemy, set up a defensive position, and again wait for orders.

The 3rd Infantry Division will be the reserve. The commanding officer of the heavy guns crew would begin a preliminary bombardment that will last for forty minutes. He will shift fire towards the second trench of the enemy while the infantry moves in to attack. He will give continuous cover fire to the two divisions.

Once the preliminary bombardment began, the division's engineers would move out to clear mines and create a passage large enough for a platoon, marking the cleared area. To the left and right sides of the cleared passage, scouts would keep guard to ensure the safety of the engineers. The day and time scheduled for the attack to begin was November 12th, 1985 at 2000 hours. Before the start of the preparatory barrage, care was to be taken so that there were not any new movements to give us away.

After getting clarification on points that were unclear, we returned to our units.

Based on the operational orders I just received, I called my brigade commanders, and the artillery commander, and gave them the following orders:

The 5th Airborne Brigade will move along the road on the left side of Hill 1702, on the right side of the defense area of the 47th Mountaineer Brigade and will eliminate the enemy. Then climb to the top of the mountain, reorganize itself, set up a defensive position against any enemy counter-attack, and then await further orders from me. The 6th Airborne Brigade will eliminate the enemy entrenched at the base of Dalicha Hill, climb to the top of the mountain, regroup, then set up a defensive position against enemy counter-attacks, and wait for my orders.

The 7th and 8th Airborne Brigade will be reserves. They will go into battle only when I give them the order to do so. I then went on to make preliminary preparations and let them know that I would be with the 5th Airborne Brigade during the battle. When I went around to look at the preparations of each brigade, I saw that they were all ardently waiting for the battle to begin. They were ready for action.

At 1920 hours, the preliminary fire began. Mortars and 130 mm heavy guns rained shells on the enemy forces in the first trench for forty minutes. The infantry, making use of the fire cover from the preparation fire, began to move from the assembly area to the assaulting line.

The 5th and 6th Airborne brigade joined the 44th Mountaineer Brigade in their trenches, to begin the attack jointly. We could see from the trenches that the front could only accommodate one battalion at a time. We revised our plan so that two battalions, one from each brigade, would attack while the remaining battalions would be held in reserve and deployed into battle when necessary, upon my order.

The assault started when the bombardment from our guns shifted from the enemy's first trench to the second one. Although the enemy suffered heavy casualties, they still put up a stiff resistance, determined not to let go of their defensive position. The 5th Airborne Brigade's first battalion followed resolutely on the heels of the battalion leading the charge. They charged courageously into the guns and bombs, stepping over the dead bodies of their brothers. Many were killed, while those fortunate enough to survive, passed the first obstacle and were poised to make history.

The 6th Airborne Brigade and the 18th Mountaineer Division brought in troops and continued their assault until the end. Our forces controlled some hills but could not move forward. A short time later, the enemy gathered its forces and launched a counter-attack on the flank and captured the hills we had recently taken under control. We reorganized and launched a counter-attack of our own but could not break through because the enemy force had been reinforced. In the meantime, our superiors issued an order for us to stop the attack and set up a strong defensive position instead.

At this point, Operation Bahre Negash was over.

On December 4th, 1985, I received a letter of appreciation from Major General Merid Negussie for my service during the third phase of the Bahre Negash Operation. The letter reads as follows:

(Confidential)

December 4,1985

To: Comrade Brig. General Tesfaye HabteMariam

Commander of the 102nd airborne division,

Hero of Socialist Ethiopia

Afabet

Re: Letter of Appreciation

Our military forces recently carried out a historic battle. This major operation, known as "Bahre Negash," involving three major battles was the first of its kind in the history of the our armed forces, and still fresh in our minds.

During this operation, the intense fighting and the actions taken by the 102nd Airborne Division including the parachute combat jump in Algena and hand to hand combat in Nakfa were truly heroic in nature. Even though the 102nd Division was comprised of new recruits, they proved to be a capable force and created a lasting fear in the hearts if our enemy.

I am proud of the bravery and professionalism with which you carried out your battle orders. The airborne division achieved their remarkable feats because of your long battlefield experience, deep knowledge of military science, and decisive leadership.

I express my sincere admiration for your swift decisions and heroic leadership in the battle we fought on the Nakfa front, especially on November 12,1985 at which time you risked your life by fighting heroically on the front lines alongside your soldiers.

Thanking you once again for your wise leadership from the training tower to the battlefield. I hope that you will be successful in the future struggles too.

Merid Negussie, Major General

The enemy, seeing that the patriotic forces had shifted from an offensive to defensive position, carried out a feint, attacking with a small number of troops to ensure that the Ethiopian forces would remain in their trenches. Then the enemy regrouped their forces and began to take Afabet and Keren under their control via Halhal. The 22nd Mountaineer Division was sent to stop them. The enemy force, however, could not be stopped with only one division. Immediately, I took the 102ndAirborne troops from Nakfa to Halhal for reinforcement. Together with the 22nd Mountaineer Division, we managed to stop them.

After a short stay in the Halhal front, I received an order that the 102nd Airborne should move to Asmara within twenty-four hours since Asmara was no vulnerable to attack. We moved from Halhal to Asmara.

While I was trying to bring about peace and security in Asmara and its surroundings, news spread that the separatists had taken the port of Zeila under their control. They used the port to receive weapons and supplies from other countries.

The responsibility of guarding the Red Sea coast and its surrounding areas from Massawa to Tio was given to the 6th Division. The orders were to travel at night from Massawa and maneuver through the surrounding area to reach Zeila at 0630 hours, to eliminate the enemy camping in the area, and take Adulis and its surrounding region under control.

The 6th Division was reinforced with one brigade from the 102nd Airborne Division to help increase the chances for success. The brigade was divided into two teams. The first team would travel at night from Massawa by a landing craft and at 0630 hours would close the road from Zeila to Tio and Assab. The second team would start from Massawa by helicopter, reach Zeila at 0630 hours, close the road heading to the western desert, and move towards the eastward toward the Red Sea, and launch its attack on the enemy. The 6th Infantry Division would push downward from the north, and approach Zeila to encircle the port city and eliminate the enemy.

The troops ordered to travel by helicopter from Massawa to Adulis, being guided by the reference map given to them at the beginning of the operation landed on the selected hill and closed all the roads leading to the western desert. The troops completed their strategic objective by taking the high grounds under their control and moving towards the Red Sea and the enemy.

The 6th Brigade was in the process of pushing toward the port of Zeila. They had orders to enter Zeila traveling by a landing craft from Massawa at night, to close the road leading from Adulis to Tito Assab, and to approach by moving the north. Due to internal separatist sympathizers and their delaying tactics, the mission did not reach Adulis on time. The Tio-Assab road remained open.

The movements of our helicopters alerted the enemy who attacked the airborne troops as they jumped down out of the helicopters. The airborne troops, however, with fixed bayonets, pulled the pins from their grenades and charged into the enemy trenches. The charge scattered the separatists, who left their wounded and equipment and fled in the direction of Tio-Assab. If the Tio-Assab road had been closed as planned, the enemy would have surrendered, or been wiped out completely.

Our failure, as always, was not in the planning and intent, but in the execution. The airborne soldiers that engaged the enemy in battle captured and sent to the back to the rear command all captured weapons, ammunition, medications, Saudi notes (riyal), important documents in Arabic and Tigrigna, boxes of biscuits and other materials.

When Operation Red Sea was over, I handed over command of the 102nd Airborne, which I had created just as a potter creates a beautiful artifact from the very Earth. After 11 years of life and death struggles on the battlefield, I was transferred to Central Command in Addis Ababa as Deputy Commander. By God's grace, I left Eritrea intact and went to my family in Addis Ababa.

Central Command was in the headquarters of the former Imperial Honor Guard, or the 1st Army Division known for its strict decorum, alertness and military professionalism.

When I arrived, morale was abysmal. Esprit de Corps and military professionalism were all but dead. Soldiers were not even given an allowance for uniforms and wore the old ones which were patched and tattered. It had been a long time since mid-level officers had received a promotion. It was an army division by name only, just an empty shell.

I recruited and trained over two thousand persons soon after I took over my position. The ground forces sent us the uniforms we requested. We distributed the uniforms and instructed them to make sure the uniforms were always ironed and worn properly.

I instituted a training curriculum so that all soldiers, regardless of rank, began to receive refresher training in tactics, use of arms and marching.

I established competition between departments and every week, there was a contest among them with prizes for those who were best in cleanliness, equipment and property maintenance, discipline, and academics. This created a healthy spirit of pride and competition. I spent my first year reviving the troop's self-confidence and pride and I made a tremendous effort to restore their former sense of self-esteem and pride in their military profession.

In 1988, when the Armed Forces Staff College began operation in Addis, I was selected as one of the first trainees. Upon graduating in 1989, I was assigned to work in the Study and Research section of the Bureau of the Chief of Staff of the Armed Forces for one year.

In 1990, I was appointed commander of the Blatain training center.

From 1990-1991, I trained and armed the 204th and 201st Division and transformed them into proper divisions and prepared them for deployment. We trained over 50,000 recruits and sent them on to supplement the various armies and corps of our ground forces.

EPRDF (Ethiopian People's Revolutionary Democratic Front) took over power while we were still training student recruits from Addis Ababa University, National Colleges and Training Centers. Our revolutionary armed forces were disbanded. My 30 years of service to the Ethiopian people, including my efforts to preserve the nation's sovereignty, was charged against me as a crime. Eventually, I was put into prison.

After 3 years in prison, I was released. Today, I live as a free man.

Appendix

In my 30 years of military service, 1961-1991 the education and training I underwent at home and abroad to develop my leadership ability and fighting capability, the positions I held and the ranks and prizes I was awarded as well as the war and battlefields in which I took part, are given in brief below:

Education at home and abroad

1961-1964, higher diploma in military science from the Haile Selassie 1st Harar military academy

1964, certificate in airborne jumping and commando training from the ground force airborne training school, Debre Zeyit.

1965-1967, diploma in special force training, from the ground force training school (Special force), Dire Dawa.

1968, certificate in training trainers, from the ground force training school, Dire Dawa.

1971, certificate in flight security coordination training, from U.S.A Texas Border academy.

1973, diploma in special force training (Green beret), from North Carolina, Fort Bragg.

1978, diploma in command and staff officer training, from Frunze academy, Soviet Union (Moscow)

1979, certificate in ideological training, from East Germany.

1988-1989, higher diploma in command and staff officer training, from the armed forces academy. Addis Ababa, Ethiopia.

Places of work and positions held

1964-1965, platoon leader, in the 3rd division, 25th battalion, in Eastern Ethiopia, Ogaden Degehabur.

1965-1967, training in special force.

1968-1975, operation and training officer and acting commander in the special force, Harareghe, Dire Dawa.

1969-1972, flight security training officer

1975, teacher in infantry school at Genet military school (Holeta)

1976, company commander in the ground force airborne battalion.

1977, company and battalion commander at Nakfa, Eritrea region.

1977, flight security commander in A.A.

1977, battalion commander in the special force.

1977-1978, commander of the first para-commando brigade, Hararghe region

1979, commander of the 506th B task force, Eritrea

1980, deputy department head in the ground force training department

1981, commander of the 18th mountaineer's division

1981-1983, commander of the airborne training school

1984-1986, commander of the 102nd airborne division and the airborne training school.

1987-1988, deputy commander of the central command

1990-1991, commander of the Blatain training center.

1991, Army commander of the Tewodros force

Battles that I took part in

1977, the battle to dislodge the 15th battalion from the encirclement by the separatists, at Nakfa, Eritrea.

1978, the Ethio-Somalia war in eastern Ethiopia.

1979, the Elabered and Meslit battle, to break the trenches around Asmara, Eritrea

1984, the Areza – Mulki – Beshoka battle, Eritrea.

1985-1987, Eritrea.

1985, counter offensive battle in Barentu.

1986, Barentu front in the Bahre Negash Operation, and the Nakfa front counter offensive.

1986, Algena front, Bahre Negash operation, Halhal front, Red Sea operation, Massawa- Zeila front.

Promotions received

1964, 2nd lieutenant

1966, Lieutenant

1972, Captain

1977, Major

1977, Lt. colonel

1981, Colonel

1985, Brigadier General

Battlefield medals awarded to me

In 1980, Medal of hero of Ethiopia, the highest possible medal in the armed forces (for unparalleled heroism.)

Medal of the wounded in the battlefield

Operation emblem of the battlefield.

Letters of Appreciation

From Colonel Kassa Gebremariam, commander of the special forces, on Oct. 1973

From Major General Merid Negussie, member of the central committee of WPE, chief administrator of the Eritrean region, and commander-in-chief of the 2nd revolutionary army, December 18, 1985, Eritrea, Asmara.

From Major General Merid Negussie, member of the central committee of WPE, chief administrator of Eritrea region and commander-in-chief of the 2nd revolutionary army, November 25, 1985, Eritrea Asmara.

Family Photos

Soon after I graduated as
2nd Lieutenant, 1958

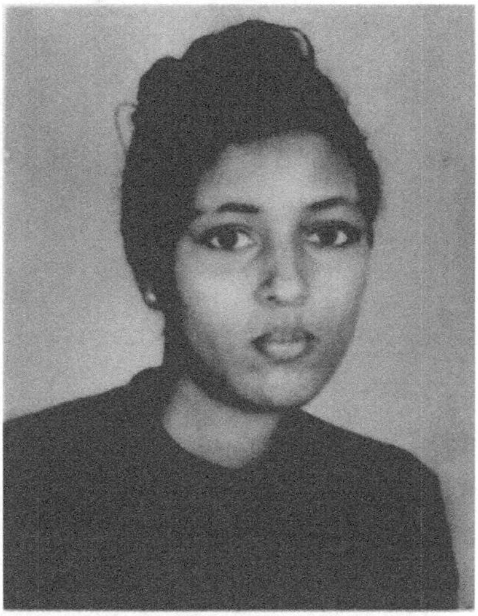

My wife, Fessessework Tessema,
in 1967

My wife and I, soon after we got married,1965

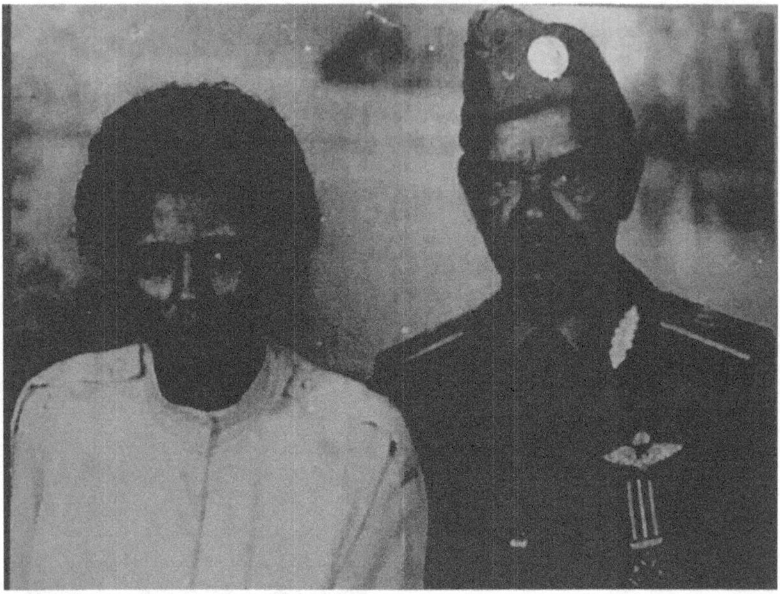

My wife and I, soon after I was promoted to the rank Brig. General, 1985.

My wife, Fessessework Tessema, and I in 2004 in Addis Ababa,
soon after my youngest daughter's wedding.

My wife and I with my 5 children and 6 grandchildren in Addis Ababa, 2005.
Back row: Awetash, Dereje, Girum, Michael, Zufan, Yodit, Woinshet, Nako, Hossana, Amare
Middle row: Fessessework, Tesfaye
Front row: Helina, Miheret, Mariamawit, Fasika

My wife and I, with my five children in Addis Ababa, in 2005.
Back row-from left to right: Awetash, Yodit, Girum, Woinshet, Amare
Front left: Fessessework, right: Tesfaye

My grandchildren in Vienna: Henok (left), Helina (Right)

My grandchildren in Ethiopia: Fikir(left), Beimnet (Right)

My surprise birthday party (78th) with some of my children and grandchildren in Virginia, USA, August, 2018.
From left to right: Gabriel, Fasika, Amare, Tesfaye, Woinshet, Nako, Hossana, Mariamawit.

Made in United States
Cleveland, OH
20 April 2026

36046722R00152